Therapeutic Landscape Research Evidence in Eco-neighbourhood Design

This book presents how to recreate therapeutic landscapes in everyday places of eco-neighbourhoods. The concept of eco-neighbourhoods goes beyond the traditional form of a residential district. Eco-neighbourhoods are characterized by many aspects related to sustainability, including protection of the environment, building social capital, ensuring a high quality of life with low economic costs, and promoting social and environmental justice. The presented work aims to systematize these phenomena and interpret them.

The action to take care of our common home, the Earth, starts locally. Creating townscapes that can promote everyday health could improve the standards of living on our planet. In both hemispheres, the majority of people live in cities; therefore, examples of good practices described in this book come from all inhabited continents. Education is the most empowering tool, which can change the future for many. Implementing eco-neighbourhoods may bring well-deserved change and hope to people in less-favoured locations of the globe.

This book will be of interest to practitioners and students of architecture, civil and environmental engineering, landscape design, spatial management, urban planning, and related fields.

Monika Trojanowska is an architect, landscape architect, and urban designer. She holds an MA diploma in Architecture and completed the postgraduate studies in Landscape Architecture. She is a university professor, licensed architect, and member of the Polish National Chamber of Architects IARP. She is the author and co-author of several dozen publications and projects in the field of architecture, urban planning, and landscape design. Her scope of interest encompasses therapeutic landscapes, parks and gardens, and eco-neighbourhood design.

Therapeutic Landscape Research Evidence in Eco-neighbourhood Design

Monika Trojanowska

LONDON AND NEW YORK

First published 2025
by Routledge
4 Park Square, Milton Park, Abingdon, Oxon OX14 4RN

and by Routledge
605 Third Avenue, New York, NY 10158

Routledge is an imprint of the Taylor & Francis Group, an informa business

British Library Cataloguing-in-Publication Data
A catalogue record for this book is available from the British Library

ISBN: 978-1-032-80001-1 (hbk)
ISBN: 978-1-032-79991-9 (pbk)
ISBN: 978-1-003-49489-8 (ebk)

DOI: 10.4324/9781003494898

Typeset in Times New Roman
by codeMantra

Contents

Introduction

The concept of therapeutic landscapes was coined by Wilbert Gesler, who defined them as places where both the physical and built environment, social conditions, and human perception together create an environment and atmosphere conducive to healing (Gesler, 1996, 2005). Thus, therapeutic landscapes are places which combine material (built environment), social, spiritual, and symbolic aspects conducive to health promotion. Therapeutic landscapes are often extraordinary places where people gather and expect healing. They have an established reputation and a documented history of spectacular, miraculous healings, like Lourdes in France, Saint Ann de Beaupré in Quebec, Canada, or Bama village in China (Gesler, 2005; Huang & Xu, 2017; Williams, 1998, 2010). However, the therapeutic experience in a given place depends on the individual perception and attitude of the user. Human perception is just as important as the material, and physical environment, so therapeutic landscapes are only potentially therapeutic places, depending on the disposition and openness of a given person. Human beings across the globe have similar basic needs and ability to experience the beauty of nature.

The concept of eco-neighbourhoods goes beyond the traditional form of residential district. Eco-neighbourhoods are characterized by many aspects related to caring for our common home, planet Earth. These include protecting the environment, building social capital, ensuring a high quality of life with low economic costs, and promoting social and environmental justice. The presented work is an attempt to systematize these phenomena and interpret them.

Why we do it?

We believe that it is possible to recreate therapeutic landscapes in everyday places. The action to take care of our common home starts locally. Caring for the health promotion in contact with nature is based on scientific research that confirmed the therapeutic qualities of landscapes. Creating townscapes that can work as everyday health-promoting places could improve the standards of living on our planet. Majority of people live in cities on both sides of the Equator. Therefore, examples of good practices described in this book come from all inhabited continents. Education is the most empowering tool, which can change the future for

DOI: 10.4324/9781003494898-1

many. Implementing eco-neighbourhoods may bring well-deserved change and hope to people in less-favoured locations of the globe.

The ultimate goal

The question behind this work was: can eco-neighbourhood become health-promoting place? The objective is to demonstrate that they can and describe the methods to improve the health-promoting qualities of everyday spaces.

Where are we now?

There are a number of assessment systems for neighbourhoods in the world. The most famous LEED, BREAM, and HQE systems have separate modules for the assessment of neighbourhoods, i.e., LEED FOR NEIGHBOURHOOD DEVELOPMENT, BREEAM COMMUNITIES, or HQE Aménagement. The French Government's ÉcoQuartier programme is a national system which is free of charges. The French ÉcoQuartier programme was transferred to other parts of the world (Colombia). However, on the global scale there are still only a few residential developments in the richest countries that apply for certification. While certificates are widely available, there are seldom any attempts at obtaining them. The most common obstacles are initial costs, extended duration of the investment, and lack of sufficient information and public awareness. Therefore, it is justified to address the issue of eco-neighbourhood design, especially all the aspects related to human health.

State of knowledge

There is a plethora of research evidence confirming beneficial influence of contact with nature on human health (WHO, 2016). Urban green spaces can promote mental and physical health to such a degree as to reduce morbidity and mortality (Maas et al., 2009). Blue and green infrastructure can alleviate the psychological stress, stimulate relaxation, as well as reduce air pollution, noise, and excessive heat. Even though there are some voices that the implications of this research for public policies may be vague due to the lack of systematic definitions of studied spaces (Lawrence, 2020), the evidence that local urban nature is related to health is growing (Boyko et al., 2021). Firstly, the greenspace improves an individual's short-term psychological health, but over a longer term the beneficial health impact may translate to greenspace reducing chronic disease and illness (Boyko et al., 2021). Gina Lovasi and her team proved the impact of the presence of trees in the city on reducing the incidence of asthma among children (Lovasi et al., 2008). Faber Taylor and Kuo (2009) found that children with ADHD syndrome could concentrate better after a walk in the park. Contact with nature promotes self-discipline (Faber Talor et al., 2002). A team of Japanese doctors led by Takehito Takano (Takano et al., 2002) proved that a well-designed system of green areas encouraging walking is a factor that directly affects the extension of the life of seniors. The presence of blue and green infrastructure can provide significant reductions in urban temperatures. Large trees can reduce local temperatures through shading and evapotranspiration. Trees can reduce air temperatures by as much as 2°C–8°C. Their presence can prevent unnecessary loss of life during heat waves (GDCI, 2024).

Greenspace within cities is an example of an ecosystem services that can be delivered. The consequences of urban development include not only biodiversity loss but also modifications in natural flows of water, treatment of used water, and the drainage of storm water (Lawrence, 2020). Thus, the grid of green and blue infrastructure is crucial for protection and restoration of natural processes to maintain the habitable conditions of our common home – the Earth. No society will maintain economic or cultural success for long if it builds it on the ashes of the natural environment. Areas with natural plant cover ensure biological continuity, retention and release of water, purification of water and air, circulation of microelements, continuity of food chains, etc. (Benyus, 2008). Open green spaces influence the retention and filtration of rainwater, prevent erosion, reduce the nuisance of heat islands in the city and lower the temperature by providing shade, reduce noise, produce oxygen, and create a living environment for many animals (Garvin et al., 2013). Pollution of the environment is not an obstacle to the survival of the human species, but it is a barrier to human development in the physical, emotional, intellectual, and spiritual spheres (Kahn, 1999).

References

Benyus J. 2008. A good place to settle: Biomimicry, biophilia, and the return of nature's inspiration to architecture w: Kellert S., Heerwagen J., Mador M. (Eds.), *Biophilic Design: The Theory, Science, and Practice of Bringing Buildings to life*. John Wiley & Sons, Inc, Hoboken, NJ, pp. 27–43.

Boyko T., Cooper R., Dunn N. 2021. *Designing Future Cities for Wellbeing*, Routledge, London.

Faber Taylor A, Kuo F.E., Sullivan W.C. 2002. Views of nature and self-discipline: *Evidence* from inner city children. *Journal of Environmental Psychology*, 22, 49–64.

Faber Taylor A, Kuo FE. 2009. Children with attention deficits concentrate better after walk in the park. *Journal of Attention Disorders*, 12, 402–409.

Garvin E.,Cannuscio C., Branas C. 2013. Greening vacant lots to reduce violent crime: A randomised controlled trial. *Injury Prevention*, 19(3), 198–203. https://doi.org/10.1136/injuryprev-2012-040439.

GDCI (Global Design Cities Initiative). 2024. *Global Street Design Guide*. National Association of City Transportation Officials, New York, available online: https://globaldesigningcities.org/publication/global-street-design-guide/resources/references, retrieved on 19.08.2024.

Gesler W. 1996. Lourdes: Healing in a place of pilgrimage. *Health & Place*, 2(2), 95–105.

Gesler W 2005. Therapeutic landscapes: An evolving theme. *Health & Place*, 11, 295–297.

Huang L., Xu H. 2017. Therapeutic landscapes and longevity: Wellness tourism in Bama. *Social Science & Medicine*, 197, 24–32. https://doi.org/10.1016/j.socscimed.2017.11.052.

Kahn P.H. 1999. *The Human Relationship with Nature: Development and Culture*. The MIT Press, Cambridge, MA, London.

Lawrence R.J. 2020. *Creating Built Environments: Bridging Knowledge and Practice Divides*. Routledge, Abingdon, Oxon, New York.

Lovasi, G.S., Quinn, J.W., Neckerman, K.M., Perzanowski, M.S., Rundle, A. 2008. Children living in areas with more street trees have lower prevalence of asthma. *Journal of Epidemiology & Community Health*, 62(7), 647.

Maas J., Verheij, R.A., de Vries S., Spreeuwenberg P., Schellevis F.G., Groenewegen, P.P. 2009. Morbidity is related to a green living environment. *Journal of Epidemiology and Community Health*, 63, 967–973.

Takano T., Nakamura K., Watanabe M. 2002. Urban residential environments and senior citizens' longevity in megacity areas: The importance of walkable green spacer. *Journal of Epidemiology Community Health*, 56, 913–918.

WHO Regional Office for Europe. 2016. Urban green spaces and health, available online: https://iris.who.int/bitstream/handle/10665/345751/WHO-EURO-2016-3352-43111-60341-eng.pdf?sequence=3&isAllowed=y, retrieved on: 13.08.2024

Williams A. 1998. Therapeutic landscapes in holistic medicine. *Social Science and Medicine,* 46(9), 1193–1203. https://doi.org/10.1016/S0277-9536(97)10048-X.

Williams A. 2010. Spiritual therapeutic landscapes and healing: A case study of St. Anne de Beaupre, Quebec, Canada. *Social Science & Medicine,* 70(10), 1633–1640. https://doi.org/10.1016/j.socscimed.2010.01.012.

1 Human health

Human health

Human health and well-being is vital for our collective future, economic growth, and stability. Health is a fundamental human right, recognized in the Universal Declaration of Human Rights in 1948.

One of the Sustainable Development Goals (SDGs), SDG 3, encompasses good health and well-being.

According to Millennium Ecosystem Assessment, well-being is the outcome of five interrelated factors: basic material for a good life (including food, shelter, and goods), health, freedom of choice, good social relations, and security (Millennium Ecosystem Assessment, 2003).

Roderick Lawrence (2021) describes several factors determining health: direct pathological effects of chemicals, biological agents and radiation; the influences of climatic, geographical and physical factors; the contribution of cultural, psychological and social dimensions of daily life, built environments and infrastructure.

Built environment should actively promote well-being. In this book, we are examining the built environment and infrastructure, trying to find possibility to counteract the tendency of cities to become unhealthy places to live.

Geof Rayner and Tim Lang (2012) analysed five models of people-health-environment interrelations: sanitary-environmental, biomedical, social-behavioural, techno-economic, and ecological public health.

The sanitary-environmental model considers that the external environment is a threat to human health (hazards, vector-borne illness and disease, pollution). These threats can be counteracted by interventions preventing the outbreak of disease (clean water, healthy food, and improved hygiene). This model is applicable to developing countries.

The biomedical model idea promotes the use of medical sciences to counterbalance the threats and improve health. This model is criticized as reactive and not proactive and leading to medicalization of everyday life.

The social-behavioural model stipulates that health is a function of information and knowledge. The core concept is human agency. Ignorance and lack of social support can cause ill health. Modification of behaviour patterns can improve

DOI: 10.4324/9781003494898-2

health. This model results in health promotion programmes and social marketing. However, it may be turned into consumerism, reducing health to cognitive factors.

The techno-economic model links poor health with low income and standard of living. The prime elevators of health are economic, political, and societal development. The responsibility for poor health is shared by governments and corporations.

The ecological public health posits dependence of health on interactions of humans with the natural world. Health improvement requires macro-environmental changes. There is little role for individual effort.

The concept of therapeutic landscapes relates to ecological model (opportunities for contact with nature) and social-behavioural model (seeking health in contact with nature).

The provision of everyday health-promoting spaces in eco-neighbourhood facilitates spending time in contact with nature every day. The human agency is needed for people to use those spaces every day. Architectural and urban design can promote active living. It is up to people to make their own healthy choices.

Microclimate

Marjorie Musy, along with other researchers (2014), draws attention to the role of urban parks, green roofs, and sustainable water management (rain gardens) in creating a microclimate. A lot can be done to ensure an appropriate microclimate at the design stage: protection against noise, excessive sunlight and winds, facilitating the infiltration and evapotranspiration of rainwater, or allowing the presence of health-beneficial phytoncides produced by plants. Water, along with greenery, is the most desirable element of landscape that promotes health. The presence of water results from the development next to natural water reservoirs (river bank, lake, or sea coast), but also artificial water features (fountain, pond). The impact of the presence of water on health and well-being is very wide, from passive exposure – seeing the water surface (observing the change of colours, the gleam of the sun), listening to the sound of water (shimmering stream, rustling fountain), to active contact – touching the water (toes or feet up to complete immersion during bathing or swimming) (Trojanowska, 2020).

Air quality

Many studies demonstrated the beneficial effect of urban green on the air quality and reported impressive pollutant removal capacity of a so-called "urban forest" for different cities (McPherson et al., 1994; Nowak & Crane, 2000; Nowak et al., 2002; Tallis et al., 2011). Deciduous trees contribute very significantly to reducing the amount of pollutants in the air. The age, species, and size of the tree's crown are important (Sæbø et al., 2012). According to Zimny, a 100-year-old common beech tree can be replaced by 1,700 one-year-old trees of the same species with a crown diameter of about 1 m, or 300–400 m^2 of greenery arranged in the city with trees, shrubs, and lawns (Zimny, 2005, p. 100). As Henryk Zimny (2005) states, a plant surface (lawn, ivy, flower meadow) retains 3.6 times more dust than a smooth surface (asphalt, pavement slabs, paving stones). Gina Lovasi and her team have proven the impact of trees in the city on reducing the incidence of asthma among children (Lovasi et al., 2008).

Clearly, the amount of trees and greenery in the vicinity is perceived as factor determining better air quality. However, the air quality effect of urban vegetation is more complex than implied by such general assumptions. Despite the fact that urban trees effectively remove pollutants from the air, under certain circumstances they may induce a local increase of concentrations. Peter Vos et al. (2013) demonstrated that roadside urban vegetation may lead to increased traffic-emitted pollutant concentrations, because trees and other types of vegetation may form an aerodynamic tunnel and reduce the ventilation. The aerodynamic effect of decrease in diluting the pollutants proved to be stronger than the pollutant removal capacity of vegetation. Tom Grylls and Maarten van Reeuwijk (2022) found that the capacity of trees to remove pollutant is related to the type of emission. For street-type pollutant such as PM2.5, deposition dominates and local air quality improves. For local emission sources such as NOx on a busy road, the dispersion effects of trees become more prominent and can lead to elevated concentrations (Grylls and van Reeuwijk, 2022).

Pollution of air, water, and land is not an obstacle to the survival of the human species, but it is a barrier to human development in the physical, emotional, intellectual, and spiritual spheres (Kahn, 1999, p. 20). No society will maintain economic or cultural success for long if it builds it on the ruins of the natural environment. Air pollution is not evenly distributed, and the perceived levels of pollution relate to both physically measured levels and social influences. The higher the socio-economic status is, the higher the concern for air quality (Bickerstaff & Walker, 2001). The study by Rosemary Day (2006) proved that people are responding to the physical levels of pollution in the air. However, people were willing to accept lower air quality if it could be traded for other symbolically therapeutic resources.

In places associated with therapeutic landscapes, and with scientifically proven good air quality, people experienced air as part of that healing environment. Places like countryside and natural environments were associated with healthy environments, good air, peace and quiet, and no stress (Day, 2006).

The indoor air quality (IAQ) is of great importance for occupants' health and well-being. Noxious compounds emitted from building materials and poor ventilation can have lasting consequences on health (allergies, cancer, genetic damage, transmission of communicable diseases). IAQ is determined by multiple factors: quality of the outside air infused into the ventilation system; pollutant emissions within the building; the ventilation rate; the efficiency of filtration; and the maintenance of mechanical systems (Friedman, 2022).

Acoustic environments

Noise pollution in residential areas may significantly affect the quality of sleep and cause serious health problem. Peace and quiet are important for housing structures. Noise-emitting devices should be located as far away from living rooms and workplaces as possible. The best is to reduce noise at its source by using acoustic and anti-vibration insulation. Townhouses and apartment buildings require sound barriers in the walls and floors between the units (Friedman, 2022).

In addition, various methods for noise reductions are designed on an urban scale, like acoustic screens or earth embankments. The role of a buffer can be played by urban greenery, recreational and sports areas, and even buildings or rooms in buildings that do not have high acoustic requirements, (sports facilities, gyms, dance studios) (Trojanowska, 2020).

Jackson Wilson et al. (2016) stated that the natural (park) environment was perceived as significantly more restorative than the artificial (street) environment. The level of restorativness is related, among other factors to noise. The effect of sound buffering vegetation around the perimeter could be enhanced by regulations, minimizing the level of mechanical and human noise in the surrounding areas. The proposed solution may be to create the natural acoustic environment (Wilson et al., 2016).

The research in Denmark demonstrated that the visual environment and the atmosphere had a great influence on the perception of soundscape in open public spaces (Bjerre et al. 2017).

The sonic aesthetic is an attribute of landscape. What is sonically pleasing can elevate the spirit and help engage in therapeutic experiences. The sonic aesthetic of a given place may be responsible for preferences of users. The sound of landscape is equally important as vision. While noise management focuses on reducing disturbing and unpleasant sounds, the soundscape planning focuses on acoustic environments that are regarded positively (Prior, 2016). A study in Curitiba confirmed that landscapes with better visual quality and larger presence of vegetation resulted in a greater amount of attenuated noise (Dias de Oliveira, 2021).

Olfactory environments

The opportunity for sensory experiences is particularly significant in therapeutic landscapes (Milligan et al., 2004). Experiences of places associated with health and well-being are created by all sensuous engagements: vision, taste, touch, sound, and smell. The olfactory composition of therapeutic landscapes is a vital part of enacting a therapeutic engagement with place. Aromas, smells and scents, all set off bodily reactions (Gorman, 2017). Smells are linked to air quality, pollution, and the distribution of environmental burdens (Porteous, 1985). The good question is how to create the scentscape of eco-neighbourhoods to promote health?

Richard Gorman (2017) draws the attention to the fact that embodied relationships with specific scents could constitute therapeutic encounters with place. The smell of farms: the animal musk of livestock, the bouquet of fragrances from horticulture and agriculture, the scent of a forest may all have therapeutic effects (Gorman, 2017).

The study by Eliza Marsh and a team (2022) demonstrated that the mood is a factor determining the positive perception of pleasant scents. When exposed to pleasant scents, the individuals with more positive baseline moods rated the odours as being more pleasant than did those with lower baseline mood scores. The inherent mood may influence the degree to which smelling normatively pleasant odours may enhance mental and physical health outcomes (Marsh et al., 2022).

The unpleasant odours convey the experience of gloomy decay and health hazards. It is difficult to combat unpleasant odours using architectural tools. It is best to neutralize odours at the source. Modern solutions in the field of selective waste collection – Automatic Waste Collection Systems (AWCS), underground installations, pneumatic collection and airtight garbage containers – are intended to serve this purpose. It is important to maintain cleanliness and empty garbage containers frequently.

Protection from the sun: urban heat islands

In large cities, where built-up areas predominate, we can observe the formation of local places where the temperature is significantly higher than the surroundings. It is the result of heating of hardened surfaces and disturbed air circulation. The use of dark or non-reflective surfaces on roofs, parking lots, driveways, pedestrian paths, and other paved surfaces contributes to the heat island effect, absorbing the sun's heat, which then radiates to the surroundings. This phenomenon, called the creation of heat islands, is particularly troublesome in summer, during periods of heat waves. The heat island is linked to the accumulation of emissions of harmful chemical and physical compounds in city areas with fully hardened surfaces that absorb solar radiation (Bartosz, 2024; Kobylarczyk, 2018).

There are various methods to protect urban environment from the creation of heat islands: preserving permeable surfaces, increasing urban open green areas, planting deciduous trees with large crowns providing shade in the summer (local temperature reduction), limiting the area of paved surfaces, limiting the area of surfaces with dark coverings, and improving air circulation (use of aeration corridors). Dark flat roofs can be exchanged for cool roofs or green roofs. Creating large volumes of roofs with high albedo or vegetation requires a higher initial cost, but has an acceptable economic payback time when integrated into the overall system to maximize energy savings. Green roofs provide benefits in many ways: additional thermal insulation, storage of rainwater, retention and filtration of dust, impact on sound transmission and reflection, and improvement of the aesthetic of buildings (Zielonko-Jung & Marchwiński, 2012).

Urban greenery can significantly contribute to the modification of air parameters that are beneficial to humans. The advantage of leafy greenery of deciduous trees is that it limits access to sunlight in summer, and in heating periods, in the absence of foliage, it does not constitute a barrier to sunlight. Shadows cast by trees may help to keep buildings cool.

The microclimate is also influenced by water reservoirs, which have the ability to store heat. They heat up slower and cool down more steadily than surrounding air. The presence of water diminishes the temperature fluctuations. That effect spreads not only next to the reservoir, but also in the adjacent areas.

The architectural design strategies to protect buildings from overheating include installation of shading solutions (permanent or movable). It is important that the solutions used to limit excessive sunlight in the summer do not reduce passive heat gains in the winter. The effectiveness of shields depends on their location. Guards installed from the outside are twice as effective as those installed from the inside. The best elements are those that reflect sunlight before they touch the glazing surface, thus preventing the greenhouse effect and overheating of the interior. The exterior architectural solutions include: eaves, balconies, canopies, light breakers, external blinds, shutters, awnings, photovoltaic panels, solar control glass (diffusing, reflective, absorbent, printed, mirrored, low emission), and even glass with variable translucency). The internal sun protection is provided with curtains, shutters, blinds, and appropriately designed vegetation (Bartosz, 2024; Kobylarczyk, 2018).

Protection from the wind

When selecting a site for planning neighbourhood and every new building, wind patterns should be considered, along sun exposure, terrain conditions, and geomorphological structure. The wind's velocity and direction may be affected by landscape features such as hills, water bodies, and existing vegetation. Wind velocity and direction change in the city centre, due to influence of build transportation corridors on local winds and aerodynamic circulation (Kobylarczyk, 2018).

There are three main wind patterns worth noting. The first is that the wind's velocity picks up over long stretches of flat surfaces. The second is that the wind flows travel inland from the water during the day and switch direction at night. The third one is that mountain ranges can also influence the direction and speed of wind. As wind approaches a mountain range, it is forced upwards. During the day, air on the mountain slopes heats up and rises. At night, the process is reversed, the cooler air from the mountain slopes sinks downwards (Friedman, 2022). Sites near bodies of water tend to be susceptible to intense winds. Wind speed increases quickly with altitude. At ground level, there are obstructions, friction with the ground, and air density that slow the wind down. However, as wind gets higher, these obstructions are removed and the wind speeds up.

Prevailing wind directions are important when choosing both the orientation and spatial form of build development. Orienting buildings parallel to the direction of wind flow, and designing relatively small facades on the windward side, may help to reduce the negative effects of wind pressure. It is necessary to check whether the new development does not block the existing aeration corridors of other neighbourhoods. Proper orientation of a dwelling according to prevailing winds can optimize the benefits of this natural phenomenon with simple design consideration.

Existing and newly planted vegetation can also play a role in controlling the wind exposure. If positioned well, and forming a barrier to cold winds in the winter, it can aid in slowing down wind speed. In the summer, planted vegetation can be used to shade a building and make advantage of winds for cooling. Thus, it can reduce the need for costly mechanical ventilation. For this purpose, coniferous trees are recommended to be positioned by the north façade to reduce the north winds (northern hemisphere). The deciduous trees should be planted in the south (northern hemisphere). The recommended distance from the building is 1.5–2.5 times the height of the building. When planted too far, the vegetation may create the inverse effect and aid in the formation of wind tunnels. If planted too close, trees may obstruct the entry of natural light (Friedman, 2022).

Therapeutic landscapes

Wilbert Gesler defined therapeutic landscapes as places where both the physical and built environment, social conditions, and human perception together create an environment and atmosphere conducive to healing (Gesler, 1996, 2005). Thus, therapeutic landscapes are places which combine material (built environment), social, spiritual, and symbolic aspects conducive to health promotion. Many researchers from various fields (environmental psychology, medicine, sociology, architecture, and urban planning) have tried to describe the unique features of therapeutic landscapes. The conducted literature research allowed for the systematization. For better clarity, the features were divided, in accordance with Gesler's definition (1996), into material aspects related to the promotion of social contacts and user perception (Figure 1.1).

The natural and manmade qualities of public spaces can make it a friendly place. United with social aspects can turn it into a health-promoting place. Only when material and social aspects are combined with symbolic and spiritual aspects of user's perception, it is possible to observe the therapeutic landscapes.

European Charter on Environment and Health (WHO, 1989) recognized that every individual is entitled to "an environment conducive to the highest attainable level of health and Wellbeing". The Charter acknowledged that clean and harmonious environment in which physical, psychological, social, and aesthetic factors are all given their due importance may bring benefits to health and well-being. According to the Charter, prevention is better than cure.

A positive finding can lead to overly physically deterministic conclusions, whereas health-promoting landscapes are always related to local conditions, needs, and possibilities. What is important is not only the mutually reinforcing relationships between places, people, and meaning-making, on the one hand, but also the institutions and political processes that shape these relationships. A balanced approach to the topic of shaping everyday health promoting places is important.

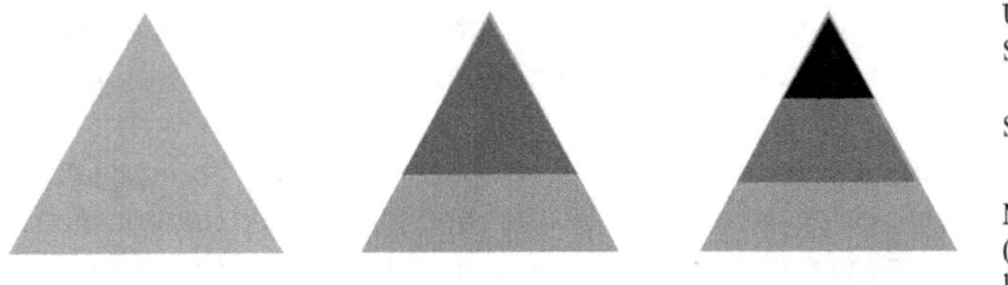

Figure 1.1 The hierarchy of various aspects of therapeutic landscapes.

Jason Corburn (2009) warns against the assumption that rational physical and urban design can change social conditions, especially for the poor. In his opinion, research on the relationship between the built environment and health sometimes fails to consider the interactions and relationships between multiple aspects: physical, social, political, economic, and spiritual, which combine to make a place (placemaking).

Places conducive to health promotion are places that combine the advantages of therapeutic landscapes to have a positive impact on physical, mental, and spiritual healing. These are places in the neighbourhood that can uplift the spirit (places of spiritual renewal, parks and squares, sports fields).

Key questions arise: What properties should an urban space have to become a therapeutic landscape? What features of urban space should be used to create a place conducive to health promotion?

Two types of natural landscapes have the greatest impact on improving health and well-being: green infrastructure (parks, gardens, forest complexes) and blue infrastructure (rivers, lakes, sea coast). Green and blue spaces are usually perceived as separate and distinct entities, standing out from the urban "grey" environment. Many residents see "green" and "blue" spaces as "therapeutic landscapes". The joy that people derive from contact with nature is explained by, among others, with the hypothesis of biophilia. The concept of biophilia was coined by Edward O. Wilson (Wilson, 1984, 2008). Biophilia is the internal bond that people feel with nature, regardless of geographic and cultural influences.

Public parks and gardens

The 2030 Agenda for Sustainable Development sets SDG 11 target 7 in (SDG 11.7) to "provide universal access to safe, inclusive and accessible, green and public spaces, in particular for women and children, older persons and persons with disabilities". As more people are aware of the benefits of contact with nature, cities create parks as plug-in urban design wherever they can. Urban dwellers need public parks of various sizes. They need large parks that may be used once a week and the small spaces provided within the vicinity of people's homes, which are often visited and used for daily recreation. Public parks are supporting an active lifestyle. Physical activity has been shown to be a major contributor to health. Sports fields and fitness equipment in parks contribute to improvement of public health, while working to reduce healthcare costs. The mere public park can be regarded as fitness machine on its own, with multiple activities available. Intuitively, greenspace is associated with physical activity and stress control (Boyko et al., 2021).

Accessible public open green spaces (POS) can help tackle health inequalities, from obesity to mental health. In residential areas and multi-use neighbourhoods, public parks and gardens can mitigate overcrowding and provide safe play areas for children (Friedman, 2022). Open public green spaces offer possibility for mental and physical regeneration. Various passive engagements can be promoted in a more tranquil atmosphere (eating, reading, playing). Social forms of recreation are also promoted in public parks. Social encounters in green areas help social inclusion and consolidate the social capital.

Today we are dealing with a process of changing the basic foundation of the presence of nature in the urban environment. Traditional methods of planning and maintaining urban greenery are giving way to a more ecological and sustainable approach. These new methods allow for natural plant succession, soil regeneration, and rainwater infiltration. Human interference is limited to a minimum. The pursuit of sustainable development and protection of biodiversity has led to the creation of a new generation of urban parks in which humans allow nature to follow its own rules. Parks of new generation are low-budget parks. Maintenance is limited to keeping the parks clean. No chemicals are used to protect plants or as fertilizers. Only plant compost is allowed to fertilize the soil. Dry leaves are often left as natural winter cover. Plant selection is limited to native species. The use of sturdy habitat plants and minimal human interference in the natural processes of ecological succession allow for reduction of park maintenance expenses. Additionally, large swathes of parks of new generation are devoted to community gardens and pedagogical gardens, which are maintained by local residents' associations. Residents are involved in the process of designing the parks of new generation from the very beginning. They decide on the selection of recreational infrastructure, evaluate projects, and put forward their own proposals for the operation of the park. Parks of new generation are intended to be an important place for the local community. They often become a community place where local residents meet and organize various events and meetings aimed at preventing social exclusion. The parks host open-air theatres, concerts, gastronomic festivals, sports competitions and other mass events (Trojanowska, 2020).

Parks of new generation are educating children and adults about sustainable development, the environment, and the human need for contact with nature. It is a place of contact with nature, mental and physical regeneration, physical activity, and social contacts.

Contact with nature

Research results indicate a positive impact of contact with nature on human health (Maas et al., 2006; Lothian, 2017). Roger Ulrich (1984) demonstrated that hospital patients recovering from surgery experienced more favourable recovery courses, including lower intake of potent narcotic pain drugs, shorter hospital stays, and better emotional states, if their windows overlooked trees rather than a brick building wall. Roger Ulrich's stress reduction theory (SRT) explains why nature exposure reduces psychological and physiological stress and enhances cognitive performance. He postulates that a central role in responding to nature is played by immediate, and unconsciously triggered affective responses (Ulrich, 2023). These reactions are inborn and innate to all humankind. Contact with nature restores involuntary attention, which, in turn, restores voluntary attention. This mechanism starts in the cognitive part of the brain. The controlled cognitive responses come after that and may vary according to cultural upbringing (Boyko et al., 2021).

Roger Ulrich's (1984) research on the impact of the view from the window on the recovery of patients after gallbladder surgery influenced the postulate that every apartment should have at least one window with a view of greenery and trees. Caring for an attractive view of greenery from the windows of rooms intended for permanent stay of people should be a priority of urban design for public health. Roger Ulrich's theory stipulates that recuperation from stress should be faster and more complete in contact with nature settings than in built environments lacking nature. Hospital patients and healthcare staff who use well-designed therapeutic gardens report improved emotional well-being and reduced stress (Ulrich, 2023).

Contact with nature enables positive health effects. Frequent contacts with nature facilitate development of connection with nature. Connection with nature is the self-perceived relationship between the self and the natural environment. Connection with nature mediates increased positive emotions. Connectedness to nature fosters spirituality and personal well-being. Connectedness to nature and spirituality explains and promotes sustainable behaviour (Navarro, 2023). Connectedness to nature is also considered an aspect of well-being (Boyko et al., 2021).

Michel Bonneti (2010) draws attention to the fact that most urban projects do not take into account the potential of the surrounding environment. They limit themselves to the boundaries of their plots. Bonneti advises to take full advantage of the opportunities offered by the landscape surroundings. In the case of revitalization of an existing urban tissue, it is worth asking whether it is possible to improve the quality of landscapes without demolishing the existing fabric, only with use of plants, elements of small architecture, urban furniture, and sculptures. Aestheticization of courtyards, squares, streets, and green areas improves the visual quality of the estate.

Foo Ah Fong (1999) proposes the traditional Japanese landscape strategy of enhancing the garden scenery by incorporating the surrounding landscape. This principle has been used in Japanese garden art for centuries to design gardens. Agata Zachariasz (2006, 2010) describes the phenomenon of shakkei (borrowed scenery) as a very important element of the composition of a Japanese and Chinese garden, used to enlarge the garden space using visual connections with the distant landscape (mountain ridge) or neighbouring elements of the scenery (trees).

Vitamin G for all

Facilitating the exposure to natural landscapes (walking, cycling, observing through a window, or looking at a picture) is related to improved mental health and well-being, restored attention, uplifted mood, better recovery from illness, reduced negative effects of stress and conditions such as depression and dementia, lower body weight (BMI), reduced overweight and obesity, as well as higher levels of physical activity, and better self-assessment of health (Balfour & Allen, 2014; Lima, 2023; Velarde et al., 2007; Ward Thompson et al., 2010). The range of specific health outcomes linked to contact with nature includes birth outcomes, cancer, cardiovascular disease, diabetes mellitus, migraines, musculoskeletal complaints, obesity, and various infectious and respiratory diseases (Boyko et al., 2021).

Lack of Vitamin G (exposure to nature) poses serious health risk, especially to children. Reduced exposure to environmental bacteria in early life, and growing up in an urban environment, are associated with developing allergies and asthma in later life. Less contact with people and animals may bring the same noxious results (Jatzlauk et al., 2017). The cultural services within urban environments may be causing humans to forget the natural world and its benefits, with knock-on ill effects to human health (James, 2018).

Limiting the time that modern children spend in natural environments is seriously affecting their cognitive development and health prospects (Louv, 2005).

Healing spaces

Contact with nature provides stimuli that allow people to focus on the present and forget about everyday problems, at least for a moment. Focusing on the "here and now" is crucial for mobilizing the body's vital forces, regeneration, and healing. Providing pleasant stimuli that engage people's attention is very important in designing public spaces.

Architecture and urban design may be "an instrument participating to cure and care" (Charras, 2023). Architecture becomes an unavoidable support with the material aspects – the natural landscape and the built environment that can have a positive impact on human health.

At the same time, it must be remembered that creating symbolic meanings is almost completely beyond the architect's capabilities. The perception of symbolism is completely subjective and is within the sphere of each person's personal experience. The field for action seems to be only generally known patterns and archetypes (religious symbols). Translating what the architect wanted to express with a given sign in space requires an additional message, because in practice it opens the field to any interpretations.

When material and social aspects of a place are combined with symbolic and spiritual qualities influencing individual perception, it is possible to observe the therapeutic landscapes and healing spaces. The consistency of the perception of the place in the symbolic and spiritual dimension with their expectations is very important for health promotion. Aspects related to perception are particularly related to a sense of responsibility and bond. Today, the problem is alienation, anonymity, and loneliness in the world of digital pseudo-relationships. Man needs his place on Earth. In our nomadic society, lost in the global village, this is especially clear.

Therapeutic gardens

Jean Stephans Kavanagh (2010) defined therapeutic garden as outdoor setting created to maximize the positive impact of contact with plants. According to American Horticultural Therapy Association, therapeutic garden is designed to facilitate passive and active interaction with nature to enable the healing processes. The American Horticultural Therapy Association discerns various sub-types of therapeutic gardens: healing gardens, enabling gardens, rehabilitation gardens, and restorative gardens. The most comprehensive definition of therapeutic gardens can be found in the Spring 2010 American Society of Landscape Architects (ASLA) Publication. The healing garden supports the overall healing process by helping patients return to full health. A meditation garden should encourage inward attention to deepen self-awareness and inner peace. A contemplative garden focuses on exploration of truths greater and beyond the individual's reach, in a contemplative, or even religious or mystical way. The term "therapeutic", however, suggests something more than healing, meditative, or contemplative. A therapeutic garden is more than any of these types of gardens (Gerlach-Springs & Healy, 2010).

Another division of therapeutic gardens based on the type of impact was proposed by Leah Diehl (2017). It is related to the therapeutic possibilities offered by the garden space. Gardens can support therapy in passive way (staying in the garden, observing garden space, listening to garden soundscape). Or it can be used as a therapeutic place itself (calming gardens). A garden can also be a place of active therapies: physical therapy, kinesitherapy, speech therapy, psychotherapy sessions, and horticultural therapy. Finally, the mere contact with nature can be a form of therapy through physical or sensory impact. No two gardens are the same, and some may have more than one purpose (Diehl, 2023). Leah Diehl introduces a vast category of "Landscapes for health", which encompasses both healing landscapes and healing gardens. She established two categories of healing gardens: enabling gardens and restorative gardens. Enabling gardens aim at improving physical function. Restorative gardens (meditation gardens, contemplative gardens, and sanctuary gardens) are places offering gentle, undemanding stimuli one can use in order to recover (Diehl, 2023).

There are a few researchers who focus on the subject of therapeutic gardens – Claire Cooper-Marcus, Marni Barnes, Naomi Sachs (1995, 1999, 2011, 2014) and Daniel Winterbottom and Amy Wagenfeld (2015). Their research, conducted in general hospitals, suggests that gardens designed in an informal, natural style with a variety of plants and flowers are more effective in reducing stress than structured or geometric gardens with hard surfaces (Cooper-Marcus & Sachs, 2014). It is believed that providing open gardens that are accessible to healthcare patients can promote stress reduction by providing views of nature, increasing control, and offering pleasant places to seek privacy or socialize (Ulrich, 1999).

There are many examples of therapeutic gardens in various types of facilities:

1 Hospices:
   Hospice and Palliative Care, Cape Cod, MA, USA
2 Nursing homes, long-term therapy centres:
   Nursing Home in Nasielsk, Poland; Nursing Home in Rabka Zdrój, Poland
3 Hospitals:
   MacKenzie Health Science Center, Edmonton, Canada; Howard Ulfelder Healing Garden, Massachusetts General Hospital in Boston, USA; Schwab Rehabilitation Hospital in Chicago, Illinois, USA; Institute of Marie Curie, Paris, France
4 Clinics, health centres, medical office complexes:

McLaren Health Care Village – Garden of Healing & Renewal, Clarkston, MI, USA; Kennedy Krieger Institute Therapy Garden in Baltimore, Maryland, USA

Another way to divide therapeutic gardens is to divide them into general or specialized, designed to treat a specific condition (Alzheimer's disease, post-traumatic stress disorder [PTSD], cancer). Known examples of specialist therapeutic gardens by type of condition are as follows:

1 Dementia, Alzheimer's disease:

   a Portland Memory Garden, Portland, USA
   b Serenada Centres, USA
   c The Sophia Louise Durbridge-Wege Living Garden in Grand Rapids, Michigan, USA
   d Graham Garden in Victoria, Canada
   e Jardin Alzheimer, CHU Nancy, Hôpital Saint Julien, France
   f Robert Doisneau Centre in Paris, France
   g Jardin de la maison de retraite spécialize "Les Aurélias" in Pollionnay, France
   h Nursing and Care Facility in Toruń, Poland

2. Therapy for PTSD:

   a Centre for War Veterans Magee Rehabilitation, Philadelphia, PA, USA
   b Military Hospital in Wrocław, Poland

3 Burntout syndrome:

   a Alnarp, Sweden
   b Nacadia Therapy Forest Garden, Danemark

4 Gardens supporting the therapy of neuroses, depressions, and other psychiatric disorders

   a Psychiatric Hospital in Gothenburg, Sweden
   b Dr. Jozef Babiński Specialist Hospital in Kraków, Poland

5 Cancer therapy

   a Maggie's Centre, Edinburgh, UK
   b Maggie's West London, UK
   c Maggie's Gartnaval, Glasgow, UK
   d Maggie's Centre, Lanarkshire, UK
   e Maggie's Centre Manchester, UK
   f Maggie's Centre, Oldham, UK
   g Maggie's, Sutton, UK
   h Hospice, Puck, Poland

6 Children's cancer

   a George Mark Children's Hospice in San Leandro, USA
   b Bonner Healing Garden, Bonner Community Hospice, Sandpoint, Idaho, USA
   c Derian House Children's Hospice, Chorley, Lancashire, UK

7 Burn therapy

   a The Legacy Oregon Burn Centre Garden, USA

8 Rehabilitation, spinal injuries

   a Buehler Enabling Garden, Chicago Botanic Gardens
   b Montecatone Rehabilitation Institute Imola, Italy
   c Horatio's Garden, Duke of Cornwall Spinal Treatment Centre in Salisbury, UK
   d Horatio's Garden, Queen Elizabeth National Spinal Injuries Unit in Glasgow, UK
   e Horatio's Garden, National Spinal Injuries Centre, Stoke Mandeville, UK
   f Horatio's Garden, Midland Centre for Spinal Injuries, Oswestry, UK
   g Horatio's Garden, Royal National Orthopaedic Hospital, Stanmore, London, UK
   h Horatio's Garden, Llandough Hospital Cardiff, Wales, UK
   i Spaulding Rehabilitation Hospital Gardens, Boston, USA

Therapeutic gardens can be a separate fragment of larger park complexes. In the case of popular public parks visited by large crowds, the therapeutic garden can be a demonstration fragment that allows the popularization of the idea of natural therapies. At the same time, it allows for a moment of rest in silence and solitude. A good example are the therapeutic gardens established in botanical gardens (Elizabeth and Nona Evans Restorative Garden, located in the Cleveland Botanical Garden in Cleveland, Ohio, USA).

Community gardens are designated areas of space that are cultivated collectively by members of a given community. They may be organized into associations or meet informally, as part of grassroots initiatives. In many places around the world, parts of public green spaces are designated as community gardens. There are two methods of organizing community gardens. The first involves allocating small, individual plots for each gardener. This system is similar to allotment gardens, but the individual plots are usually much smaller. In the second model, all interested parties work together according to the established schedule. The second model allows everyone willing to work to join immediately, without having to wait for their plot. There are various interesting programmes around the globe (Mujeres que Reverdecen in Colombia, Paris Jardine, France).

The pedagogical gardens for children are established in kindergartens, schools, and cultural centres. Their mission is to teach children how to grow their own food and establish a healthy connection with nature, promote biodiversity, and protect wildlife. Children are also learning to regulate their emotional states, find joy and restoration in contact with nature.

Everyday health-promoting places

A beautiful living environment increases the quality of life. Scenic beauty is a common good. It is worth considering whether the part of the area with the most beautiful view should become a public park. The practice of allocating wasteland, wetlands, swampy areas, etc. that are unsuitable or too difficult to build on, as recreational areas is not effective. Today, such areas are designated as pieces of green infrastructure excluded from human use. The beauty of the landscape can have a therapeutic effect on humans and should be used to promote public health.

The beauty of the surrounding landscape influences a person's well-being. The quality of life also depends on the visual quality of our surroundings. The space between buildings influences the perception of the quality of the urban environment. It builds a sense of identity and influences the development of an attitude of co-responsibility for the surrounding space. Shaping a sustainable environment of urban settlements requires a combination of pro-ecological solutions with architecture and urban planning of high aesthetic quality. Karmanov and Hamel (2008) argue that attractive and well-designed urban environment can reduce the negative effects of stress and stimulate our moods just like a beautiful natural environment. Christopher Alexander (2002, 2008) created a theory of natural and man-made beauty contained in 15 points – properties. The 15 properties are: levels of scale, strong centres, boundaries, alternating repetition, positive space, good shape, local symmetries, deep interlock and ambiguity, contrast, gradients, roughness, echoes, the void, simplicity and inner calm, and not separateness (Alexander, 2002, 2008).

Not every property is found in every beautiful place, but you will usually notice many of them in very beautiful buildings and facilities. These properties can be applied to the physical design and environmental construction of health-promoting landscapes of everyday life.

Gesler (1996) defined therapeutic landscapes as places where physical and built environments, social conditions, and human perceptions combine to produce an atmosphere that is conducive to healing. Erwin Zube (1987) argues that individual perceptions and responses to landscapes are shaped by experience, personal utility functions, and social and cultural contexts. Thus, individual responses to therapeutic landscapes may vary. The landscape is "a product of the human mind, and of material circumstances" (Williams, 1998 & 2010). The same space may be perceived as therapeutic by one person and as unsettling by another (Bell et al., 2017).

The designation therapeutic landscape is usually reserved to specific places of established salutogenic reputation. The description of health-affirming landscapes may be more extensive, as it refers to more common places that unite the qualities of therapeutic landscapes to influence people's physical, mental, and spiritual healing (Trojanowska & Sas-Bojarska, 2018). Both health-promoting places and therapeutic landscapes share major therapeutic attributes.

The most fascinating question is what are the therapeutic and health-promoting attributes? This long-term study began over 20 years ago, in 2001, and is still continued. The aim is to systematize the qualities of therapeutic landscapes. The study encompassed visits in over 150 public open green spaces (POS) and private therapeutic gardens in Europe and the United States. The list includes well-known parks in New York and Paris, as well as several parks in Poland. The aim was to visit and assess not only famous parks, but also less-known spaces. The condition was that they were referred to as favourite places of recreation by local residents.

The human perception was treated as a social proof of the therapeutic qualities of a landscape. The iterative process conducted for over 20 years led to development of a conceptual framework for a universal standard for health-promoting urban places.

The final draft for the standard (2024) is presented in Table 1.1. The qualities were divided into five categories: (1) sustainability, (2) accessibility, (3) amenities, (4) design, and (5) placemaking. Those categories were used to organize the qualities of therapeutic landscapes in a legible manner. The universal standard is a simple and effective tool that can be used by both professional designers and non-professionals to improve the health-promoting qualities of POS.

Proposed methodology of assessment with the universal standard for health-promoting urban places

Each of the five categories includes sub-categories and individual attributes. The universal standard for health-promoting urban places can be used for binary or detailed assessment. The binary assessment has three categories (0, ½, and 1):

0 – No, not observed
½ – Partially observed, but there is still room for improvement
1 – Yes, satisfactory

The maximum number of points for binary assessment is presented in Table 1.2. Simple manual calculation method was used to add the points. A customized Excel spreadsheet was used to verify the results.

The detailed assessment required a written explanation of why the researcher thought that the attribute was present, satisfactory, and worthy granting a point.

Proposed methodology for assessment of individual categories is provided in Tables 1.3–1.7. For better clarity, the results of the assessment are grouped into five tables (Tables 1.3–1.7). Each of the tables provides general description of requirements and lists the maximum number of points to obtain in each category for every individual feature.

The standard consists of five categories:

1 Sustainability

   This section includes all the aspects relating to the sustainable design of eco-neighbourhoods and public open green spaces (POS) as everyday health-promoting spaces (Table 1.3).

2 Accessibility

   This category is the walkability assessment. Universal accessibility, understood as addressing the needs of people with special needs, is directly linked to the promotion of health. The quality of the pedestrian routes leading to the public open green spaces is as important as the design of public parks (Table 1.4).

3 Amenities

   This section includes salutogenic features and sports and recreational infrastructure to satisfy all basic needs of users (Table 1.5).

Table 1.1 Standard for health-promoting urban places

1. Sustainability	*2. Accessibility*	*3. Amenities*	*4. Design*	*5. Placemaking*
1.1 Environmental characteristics	**2.1 Distance to park**	**3.1 Catering to basic needs**	**4.1 Urban design**	**5.1 Social engagement**
Area	**2.2 Sidewalk Infrastructure**	Places to sit and rest/benches	Architectural variety of urban environment	Personalizing the architectural process
Location	Width of sidewalk	Shelter	Quality of connections with surrounding urban pattern	Public Participation
Soil quality	Evenness of surface	Restrooms	Edges	Determining the rules of conduct and self-management
Water quality	Lack of obstructions	Drinking water/fountain	Orderly plan	Organization of events
Air quality	Slope	Food (possibility to buy food in park or nearby)	**4.2 Architectural design**	Personalization
Noise level	Sufficient drainage	**3.2 Psychological and physical regeneration**	Legible structure of paths	Animation
Green infrastructure	**2.3 General conditions of walkways**	Natural landscapes	Pockets of activity	Space for social contacts:
Blue infrastructure	Maintenance	Open green space	Human scale	– third places
1.2 New generation of public parks	Overall aesthetics	Meadow	Structure of interior connections	– fourth places
1.3 Forms of natural protection	Street art	Urban forest	Framed views	**5.2 Promotion of social contacts**
1.4 Biodiversity protection	Sufficient seating	Water feature	Long vistas (extent)	Tot lot
Parts of open green space not available to visitors	Perceived safety	Brine graduation towers	Pathways with views	Playground
Native plants	Buffering from traffic	Labyrinth	Invisible parts of the scenery (vistas that engage the imagination)	Splash pad
Native animals	Street activities	Places to rest in the sun and in the shade	Possibility to watch other people	Meeting places for groups
Natural maintenance methods	Vacant lots	Places to rest in quiet and solitude	Possibility to see wildlife	Recreational square
No invasive species	**2.4 Traffic**	Meditation garden		Picnic Tables
	Speed	Relax/Chill-out zone		Pavilion/Gazebo with running water and electricity
	Volume	**3.3 Promotion of physical activities**		Garden grill
	Number and safety of crossings	Addressing the needs of people with special needs,		
	Traffic calming	Path with a springy, shock-absorbing surface		
	On-street parking	Trail		

(*Continued*)

Table 1.1 (Continued)

1. Sustainability	*2. Accessibility*	*3. Amenities*	*4. Design*	*5. Placemaking*
1.5 Sustainable water management	**2.5 User Experience**	Bicycle path	**4.3 Salutogenic design**	Place for a bonfire/Fire pit
Rainwater infiltration	Air quality	Chess tables	Focal points and landmarks	Community gardens
Irrigation with non-potable water	Noise level	Street workout	Optimal level of complexity	Open-air stage
Stormwater retention	Sufficient lighting	Parkour	Engaging features	Off-leash dog park
1.6 Urban metabolism	Sunshine and shade	Trampoline	Controlled risk	Dog waste pickup bags
Waste segregation and collection	Visibility of nearby buildings	Rope park	Mystery/Fascination	**5.3 Human perception – spiritual and symbolic**
Bio-composting	**2.6 Bicycle infrastructure**	**3.4 Sports infrastructure**	Movement	Sacred places
Phytoremediation	Separate paths for bicycles	Sport field (soccer/football)	**4.4 Sensory stimuli design**	Shrine/Chapel
Hydrophyte sewage treatment plant	Repair stations	Street sports field	Sight stimuli	Rosary garden
1.7 Ecological energy sources	Showers & lockers	Baseball field	Hearing stimuli	Station of the cross
Solar panels	Parking for bicycles	Basketball court	Smell stimuli	**5.4 Culture and connections to the past**
PV panels	**2.7 Public transport stops**	Volleyball court	Touch stimuli	Monuments
Biomass	Bus	Tennis courts	Taste stimuli	Memorials
Biogas	Urban rail transit	Running loop	Sensory path	Historic places
Biofuel	Train	Outdoor gym	**4.5 Soundscape**	Educational features
Micro co-generation	Trolleybus	Fitness stations	Sounds of water	**5.5 Art**
Micro wind turbines	Other	Boules track	Sounds of plants (grasses)	Painting
Micro grid	**2.8 Sufficient Parking**	Tennis table	Local wild birds	Sculpture
Self-sufficiency	**2.9 Safety and security**	Teqball table	Sound installations	Architecture
Other	Presence of guards	Badminton court	**4.6 Thematic gardens**	Music
	Monitoring	Climbing wall	Biblical garden	Theatre
	Emergency devices	Skate park	Memory garden	Dance
		Pump track	Sensory garden	Literature
		Running track	Rosarium	
		Lake	Herbarium	
		Swimming pool	Medicinal plants	
		Skating ring/ice ring	Edible garden	
		Toboggan hill	Educational garden	
		Ski track		

Source: Author, improved version of the standard (2024)

Table 1.2 Maximum number of points for binary assessment

	Maximum number of points
Total	**370**
Categories	
1 Sustainability	**208**
1.1 Environmental characteristics	8
1.2 New generation of public parks	1
1.3 Legal protection of nature/nature conservation	60
1.4 Biodiversity protection	122
1.5 Sustainable water management	3
1.6 Urban metabolism	4
1.7 Ecological energy sources	10
2 Accessibility	**41**
2.1 Distance to park	1
2.2 Sidewalk Infrastructure	5
2.3 General conditions of walkways	8
2.4 Traffic	5
2.5 User experience	5
2.6 Bicycle infrastructure	4
2.7 Public transport stops	9
2.8 Sufficient parking	1
2.9 Safety & security	3
3 Amenities	**47**
3.1 Catering for basic needs	5
3.2 Psychological and physical regeneration	11
3.3 Promotion of physical activities	9
3.4 Sports infrastructure	22
4 Design	**37**
4.1 Urban design	4
4.2 Architectural design	10
4.3 Salutogenic design	6
4.4 Sensory stimuli design	6
4.5 Soundscape	4
4.6 Thematic gardens	7
5 Placemaking	**37**
5.1 Social engagement	9
5.2 Promotion of social contacts	13
5.3 Human perception – spiritual & symbolic	4
5.4 Culture & connections to the past	4
5.5 Art	7

Source: Author

Table 1.3 Assessment of sustainability (right column – number of points)

1. Sustainability		208
1.1 Environmental characteristics		**8/8**
Area (points are attributed if the public park of certain category has sufficient area)	Pocket park < 2 ha Neighbourhood park 2–10 ha City park > 10 ha Large scale park > 50 ha	1
Location	Proximity to users (residential area, mixed-use area)	1
Soil quality	Sufficient for recreational use; No visible traces of pollution	1
Water quality	Sufficient for recreational use; No visible traces of pollution	1
Air quality	Sufficient for recreational use. No traces of pollution	1
Noise level	No nuisance to moderate noise nuisance	1
Green infrastructure	Does the park form part of the green infrastructure? If YES – 1 point	1
Blue infrastructure	Does the park form part of the blue infrastructure? If YES – 1 point	1
1.2 New generation of public parks		**1/1**
New generation of public parks	Can the park be considered as a park of new generation? If YES – 1 point	1
1.3 Legal protection of nature/nature conservation		**Max 60**
Protected species	One point for every protected species, up to 20 points	Max 20
Area of natural conservation (reserve)	One point for every area of natural conservation (reserve), up to 20 points. If the public park encompasses one area of natural conservation (reserve) – 1 point	Max 20
Area of landscape protection	One point for every area of landscape protection up to 20 points. If the public park forms part of landscape protection park – 1 point	Max 20
1.4 Biodiversity protection		**Max 122**
Parts of open green space not available to visitors	Are there any secluded areas for biodiversity protection?	1
Native plants	Planting scheme is based on native species. One point for every native specie of plants present, up to 100 points	Max 100
Native animals	Are there native species of animals present in the park? One point for every native specie of animals present, up to 20 points	Max 20
Natural maintenance methods	What kind of methods of maintenance is used? If only natural – 1 point	1

(*Continued*)

Table 1.3 (Continued)

1. Sustainability		208
1.5 Sustainable water management		**3**
Rainwater infiltration	Porous, permeable surfaces	1
Irrigation with non-potable water	If irrigation with non-potable water is used – 1 point	1
Stormwater retention	Are there sustainable urban drainage systems (SUDS) used? If YES – 1 point	1
1.6 Urban metabolism		**4**
Waste segregation & collection	Is waste segregation facilitated? If YES – 1 point	1
Bio-composting	Is bio-composting facilitated? Are there bio-composting stations (At least one)? If YES – 1 point	1
Phytoremediation	Are there installations for phytoremediation? If YES – 1 point	1
Hydrophyte sewage treatment plant	Is there a hydrophyte sewage treatment plant? If YES – 1 point	1
1.7 Ecological energy sources		**10**
Solar panels	Are there solar panels used in the neighbourhood? If YES – 1 point	1
PV panels	Are there PV panels used in the neighbourhood? If YES – 1 point	1
Biomass	Are there biomass installations used in the neighbourhood? If YES – 1 point	1
Biogas	Are there biogas installations used in the neighbourhood? If YES – 1 point	1
Biofuel	Are there biofuel installations used in the neighbourhood? If YES – 1 point	1
Micro co-generation	Are there micro co-generation installations used in the neighbourhood? If YES – 1 point	1
Micro wind turbines	Are there micro wind turbines used in the neighbourhood? If YES – 1 point	1
Microgrid	Is the microgrid used in the neighbourhood? If YES – 1 point	1
Self-sufficiency	Is the neighbourhood energy self-sufficient? If YES – 1 point	1
Other	Are there any other ecological energy sources solutions used in the neighbourhood? If YES – 1 point	1

Source: Author

4 Design

This section encompasses the distribution of functions within the park space and the organization of its grid of connections (Table 1.6). It is important that the design of a park is legible and comprehensible. The composition should be harmonious. Some attributes are important when it comes to engaging the interest of users, such as mystery, risk, and movement. A separate category is dedicated to multi-sensory stimuli and sensory paths.

Table 1.4 Assessment of accessibility (right column – number of points)

2 Accessibility		41
2.1 Distance to park (Is it possible to walk to park?) If YES – 1 point		1/1
Pocket park < 2 ha	Majority of users live/work < 200 m (7 min walk)	
Neighbourhood park 2–10 ha	Majority of users live/work 200–500 m (15 min walk)	1
City park > 10 ha	Majority of users live/work 200–1,000 m (15 min walk bike/drive/public transport)	
Large scale park > 50 ha	Majority of users live/work 200–4,000 m (15–30 min walk/ bike/drive/public transport)	
2.2 Sidewalk infrastructure		**5**
Width of sidewalk	The width of sidewalk is sufficient for active use. If YES – 1 point	1
Evenness of surface	The surface is even, no thresholds. If YES – 1 point	1
Lack of obstructions	No obstruction, clear passage. If YES – 1 point	1
Slope	The sidewalks are flat and accessible to people in wheelchair (<8% slope). If the slope is > 6% and the ramp is longer than 9 m, please provide flat space for rest 1.5 m × 1.5 m, every 9 m.	1
Sufficient drainage	Sufficient for walking, no stagnating water. If YES – 1 point	1
2.3 General conditions of walkways		**8**
Maintenance	Is the neighbourhood regularly cleaned (at least once daily)? If YES – 1 point	1
Overall aesthetics	Is the neighbourhood aesthetically pleasing (general impression)? If YES – 1 point	1
Street art	Are there any artistic installations, and no signs of abuse, graffiti, etc.? If YES – 1 point	1
Sufficient seating	Are three benches with armrest and backrest every 200 m along pathways? If YES – 1 point	1
Perceived safety	Are the walkways perceived as safe places (general impression)? If YES – 1 point	1
Buffering from traffic	Is there a physical barrier between walkways and traffic (parked cars, planters with greenery, trees, bollards) (general impression)? If YES – 1 point	1
Street activities	Are there occasional events, organized or spontaneous? If YES – 1 point	1
Vacant lots	Are there any vacant lots adjacent to park? If NO – 1 point	1
2.4 Traffic		**5**
Speed	Is there a speed limit to promote walkability? If YES – 1 point	1
Volume	Are there solutions to minimize the volume of traffic? (detours, alternative roads, walls, barriers)? If YES – 1 point	1
Number and safety of crossings	Are there safe crossing next to open public green spaces? Are there numerous possibilities for safe crossing of the street? If YES – 1 point	1
Traffic calming	Are there solutions to calm down the traffic (speed bumps, stop signs, traffic signs, woonerfs, etc.)? If YES – 1 point	1
On-street parking	Is there space for on-street parking? If YES – 1 point	1

(*Continued*)

Table 1.2 (Continued)

2 Accessibility		41
2.5 User experience		**5**
Air quality	The air quality in the neighbourhood is at least good (according to Air Quality Index[1]). If YES – 1 point	1
Noise level	The noise level is perceived as moderate. If YES – 1 point	1
Sufficient lighting	Are there sufficient numbers of lamps? If YES – 1 point	1
Sunshine and shade	Are there trees or other structures to provide shade? Are there places to rest in sunshine? If YES – 1 point	1
Visibility of nearby buildings	Is it possible to see the buildings located nearby? If YES – 1 point	1
2.6 Bicycle infrastructure		**4**
Separate paths for bicycles	Is there a separate path for bicycles? If YES – 1 point	1
Repair stations	Are there any repair stations? If YES – 1 point	1
Showers & lockers	Are there any showers & lockers? Are they sufficient in number and quality? If YES – 1 point	1
Parking for bicycles	Are there any parking for bicycles? Is it sufficient in number and quality? If YES – 1 point	1
2.7 Public transport stops		**9**
Bus	There are bus stops near the park (<15 min walk).	1
Urban rail transit	There are urban rail transit stops near the park (<15 min walk).	1
Train	There are train stops near the park (<15 min walk).	1
trolleybus	There are trolleybus stops near the park (<15 min walk).	1
other	There are other public transport stops near the park (<15 min walk).	MAX 5
2.8 Sufficient parking		**1**
	Yes, there are numerous parking spots in the park. One can always find a parking spot close to park entrances (<15 min walk).	1
2.9 Safety and security		**3**
Presence of guards	Are there guards patrolling the neighbourhood? If YES – 1 point	1
Monitoring	Is there monitoring? If YES – 1 point	1
Emergency devices	Are there emergency devices to call help if needed? If YES – 1 point	1

Source: Author

5 Placemaking

Public parks offer an ideal possibility for various kinds of social contacts for people from usually isolated and disadvantaged social groups (the elderly, people with special needs). Placemaking is related to the popularity of a park. This section combines attributes that are related to the promotion of moderate social contacts and human perceptions (Table 1.7).

Table 1.5 Assessment of amenities (right column – number of points)

3 Amenities		47
3.1 Catering to basic needs		**5**
Places to sit and rest/ benches	Are there any? Are they satisfactory? If YES – 1 point	1
Shelter	Are there any? Are they satisfactory? If YES – 1 point	1
Restrooms	Are there any? Are they satisfactory? If YES – 1 point	1
Drinking water/fountain	Are there any? Are they satisfactory? If YES – 1 point	1
Food (possibility to buy food in park or nearby)	Are there any? Are they satisfactory? If YES – 1 point	1
3.2 Psychological and physical regeneration		**11**
Natural landscapes	Are there places which give an impression of a pristine natural landscape? If YES – 1 point	1
Open green space	Are there any open spaces? If YES – 1 point	1
Meadow	Are there any? Are they satisfactory? If YES – 1 point	1
Urban forest	Are there any? Are they satisfactory? If YES – 1 point	1
Water feature	Are there any water features in the park? If yes, in what form? If YES – 1 point	1
Brine graduation towers	Are there any? Are they satisfactory? If YES – 1 point	1
Labyrinth	Are there any? Are they satisfactory? If YES – 1 point	1
Places to rest in the sun and shadow	Are there any places to sit and rest in the sun and shadow? Are they satisfactory? If YES – 1 point	1
Places to rest in silence and solitude	Are there any benches to rest and enjoy silence and solitude? Are they satisfactory? If YES – 1 point	1
Meditation garden	Are there any? Are they satisfactory? If YES – 1 point	1
Relax/Chill-out zone	Are there any? Are they satisfactory? If YES – 1 point	1
3.3 Promotion of physical activities		**9**
Addressing the needs of people with special needs	Is the park area accessible? If YES – 1 point	1
Path with a springy, shock-absorbing surface	Are there any? Are they satisfactory? If YES – 1 point	1
Trail	Are there any? Are they satisfactory? If YES – 1 point	1
Bicycle path	Are there any? Are they satisfactory? If YES – 1 point	1
Chess tables	Are there any? Are they satisfactory? If YES – 1 point	1
Street workout	Are there any? Are they satisfactory? If YES – 1 point	1
Parkour	Are there any? Are they satisfactory? If YES – 1 point	1
Trampoline	Are there any? Are they satisfactory? If YES – 1 point	1
Rope park	Are there any? Are they satisfactory? If YES – 1 point	1
3.4 Sports infrastructure		**22**
Sport field (Soccer/ Football)	Are there any? Are they satisfactory? If YES – 1 point	1
Street sports field	Are there any? Are they satisfactory? If YES – 1 point	1
Baseball field	Are there any? Are they satisfactory? If YES – 1 point	1
Basketball court	Are there any? Are they satisfactory? If YES – 1 point	1
Volleyball court	Are there any? Are they satisfactory? If YES – 1 point	1
Tennis courts	Are there any? Are they satisfactory? If YES – 1 point	1
Running loop	Are there any? Are they satisfactory? If YES – 1 point	1

(*Continued*)

Table 1.5 (Continued)

3 Amenities		47
Outdoor gym	Are there any? Are they satisfactory? If YES – 1 point	1
Fitness Stations	Are there any? Are they satisfactory? If YES – 1 point	1
Boules track	Are there any? Are they satisfactory? If YES – 1 point	1
Tennis table	Are there any? Are they satisfactory? If YES – 1 point	1
Teqball table	Are there any? Are they satisfactory? If YES – 1 point	1
Badminton court	Are there any? Are they satisfactory? If YES – 1 point	1
Climbing wall	Are there any? Are they satisfactory? If YES – 1 point	1
Skate park	Are there any? Are they satisfactory? If YES – 1 point	1
Pump track	Are there any? Are they satisfactory? If YES – 1 point	1
Running track	Are there any? Are they satisfactory? If YES – 1 point	1
Lake	Are there any? Are they satisfactory? If YES – 1 point	1
Swimming pool	Are there any? Are they satisfactory? If YES – 1 point	1
Skating ring/ice ring	Are there any? Are they satisfactory? If YES – 1 point	1
Toboggan hill	Are there any? Are they satisfactory? If YES – 1 point	1
Ski track	Are there any? Are they satisfactory? If YES – 1 point	1

Source: Author

Table 1.6 Assessment of design (right column – number of points)

4 Design		37
4.1 Urban design		4
Architectural variety of urban environment	Is the Architectural variety observed in the surrounding urban environment? If YES – 1 point	1
Quality of connections with surrounding urban pattern	Are the connections with surrounding urban pattern of high quality? If YES – 1 point	1
Edges	Are the edges well defined?	1
Orderly plan	Is the plan legible and orderly?	1
4.2 Architectural design		**10**
Legible structure of paths	Is the structure and spatial organization of paths legible? If YES – 1 point	1
Pockets of activity	Are there any? Are they satisfactory? If YES – 1 point	1
Human scale	Is the design respecting the human scale? If YES – 1 point	1
Structure of interior connections	Is there a clear structure of interior connections? If YES – 1 point	1
Framed views	Are there any? Are they satisfactory? If YES – 1 point	1
Long vistas (extent)	Are there any? Are they satisfactory? If YES – 1 point	1
Pathways with views	Are there any? Are they satisfactory? If YES – 1 point	1
Invisible parts of the scenery (vistas that engage the imagination)	Are there any? Are they satisfactory? If YES – 1 point	1
Possibility to watch other people	Are there any possibilities to observe other people? Are they satisfactory? If YES – 1 point	1
Possibility to see wildlife	Are there any possibilities to observe animals? Are they satisfactory? If YES – 1 point	1

(*Continued*)

Table 1.6 (Continued)

4 Design		37
4.3 Salutogenic design		**6**
Focal points and landmarks	Are there any clear landmarks? If YES – 1 point	1
Optimal levels of complexity	Are there any features in this category? Are they satisfactory? If YES – 1 point	1
Engaging features	Are there any features in this category? Are they satisfactory? If YES – 1 point	1
Controlled risk	Are there any features in this category? Are they satisfactory? If YES – 1 point	1
Mystery/fascination	Are there any features in this category? Are they satisfactory? If YES – 1 point	1
Movement	Are there any features in this category? Are they satisfactory? If YES – 1 point	1
4.4 Sensory stimuli design		**6**
Sight stimuli	Are there any features in this category? Are they satisfactory? If YES – 1 point	1
Hearing stimuli	Are there any features in this category? Are they satisfactory? If YES – 1 point	1
Smell stimuli	Are there any features in this category? Are they satisfactory? If YES – 1 point	1
Touch stimuli	Are there any features in this category? Are they satisfactory? If YES – 1 point	1
Taste stimuli	Are there any features in this category? Are they satisfactory? If YES – 1 point	1
Sensory path	Are there any features in this category? Are they satisfactory? If YES – 1 point	1
4.5 Soundscape		**4**
Sounds of water	Are there any features in this category? Are they satisfactory? If YES – 1 point	1
Sounds of plants (grasses)	Are there any features in this category? Are they satisfactory? If YES – 1 point	1
Local wild birds	Are there any features in this category? Are they satisfactory? If YES – 1 point	1
Sound installations	Are there any features in this category? Are they satisfactory? If YES – 1 point	1
4.6 Thematic gardens		**7**
Biblical garden	Are there any features in this category? Are they satisfactory? If YES – 1 point	1
Memory garden	Are there any features in this category? Are they satisfactory? If YES – 1 point	1
Sensory garden	Are there any features in this category? Are they satisfactory? If YES – 1 point	1

(*Continued*)

Table 1.6 (Continued)

4 Design		37
Rosarium	Are there any features in this category? Are they satisfactory? If YES – 1 point	1
Herbarium	Are there any features in this category? Are they satisfactory? If YES – 1 point	1
Medicinal plants	Are there any features in this category? Are they satisfactory? If YES – 1 point	1
Edible garden	Are there any features in this category? Are they satisfactory? If YES – 1 point	1

Source: Author

Table 1.7 Assessment of placemaking (right column – number of points)

5. Placemaking		37
5.1 Social engagement		**9**
Personalising the architectural process	Are the persons responsible for the design, construction and management known to the public? If YES – 1 point	1
Public participation	Do all stakeholders, including inhabitants and users have a real influence on the design and maintenance through participatory process? If YES – 1 point	1
Determining the rules of conduct and self-management	Are there any rules of conduct and self-management established and publicly available? Are there any information boards with rules of conduct and self-management placed in the public space? Were the rules of conduct and self-management established in a participatory process? Do all stakeholders including inhabitants and users agree upon the common rules of conduct and self-management? If YES – 1 point	1
Organization of events	Are there any events organized? Are they popular/ frequented? If YES – 1 point	1
Personalization	Are there any features in this category? Are they satisfactory? If YES – 1 point	1
Animation	Are there any features in this category? Are they satisfactory? If YES – 1 point	1
Space for social contacts	Is there an inclusive, accessible space for social contacts available to all? If YES – 1 point	1
- third places	Can it be the third place for anyone? If YES – 1 point	1
- fourth places	Can it be the fourth place for anyone? If YES – 1 point	1
5.2 Promotion of social contacts		**13**
Tot lot	Are there any features in this category? Are they satisfactory? If YES – 1 point	1
Playground	Are there any features in this category? Are they satisfactory? If YES – 1 point	1
Splash pad	Are there any features in this category? Are they satisfactory? If YES – 1 point	1

(*Continued*)

Table 1.7 (Continued)

5. Placemaking		37
Meeting places for groups	Are there any features in this category? Are they satisfactory? If YES – 1 point	1
Recreational square	Are there any features in this category? Are they satisfactory? If YES – 1 point	1
Picnic tables	Are there any features in this category? Are they satisfactory? If YES – 1 point	1
Pavilion/Gazebo with running water and electricity	Are there any features in this category? Are they satisfactory? If YES – 1 point	1
Garden grill	Are there any features in this category? Are they satisfactory? If YES – 1 point	1
Place for a bonfire/ Fire pit	Are there any features in this category? Are they satisfactory? If YES – 1 point	1
Community gardens	Are there any features in this category? Are they satisfactory? If YES – 1 point	1
Open-air stage	Are there any features in this category? Are they satisfactory? If YES – 1 point	1
Off-leash dog park	Are there any features in this category? Are they satisfactory? If YES – 1 point	1
Dog waste pickup bags	Are there any features in this category? Are they satisfactory? If YES – 1 point	1
5.3 Human perception – spiritual & symbolic		4
Sacred places	Are there any features in this category? Are they satisfactory? If YES – 1 point	1
Shrine/Chapel	Are there any features in this category? Are they satisfactory? If YES – 1 point	1
Rosary garden	Are there any features in this category? Are they satisfactory? If YES – 1 point	1
Station of the cross	Are there any features in this category? Are they satisfactory? If YES – 1 point	1
5.4 Culture and connections to the past		4
Monuments	Are there any features in this category? Are they satisfactory? If YES – 1 point	1
Memorials	Are there any features in this category? Are they satisfactory? If YES – 1 point	1
Historic places	Are there any features in this category? Are they satisfactory? If YES – 1 point	1
Educational features	Are there any features in this category? Are they satisfactory? If YES – 1 point	1
5.5 Art		7
Painting	Are there any features in this category? Are they satisfactory? If YES – 1 point	1
Sculpture	Are there any features in this category? Are they satisfactory? If YES – 1 point	1

(*Continued*)

Table 1.7 (Continued)

5. Placemaking		37
Architecture	Are there any features in this category? Are they satisfactory? If YES – 1 point	1
Music	Are there any features in this category? Are they satisfactory? If YES – 1 point	1
Theatre	Are there any features in this category? Are they satisfactory? If YES – 1 point	1
Dance	Are there any features in this category? Are they satisfactory? If YES – 1 point	1
Literature	Are there any features in this category? Are they satisfactory? If YES – 1 point	1

Source: Author

Life quality

Living conditions are often determined by four basic elements:

- economic well-being
- living conditions (number of households per apartment; number of people per room; apartment equipped with heating, bathroom)
- social infrastructure (public safety, health services, medical care)
- environment in which humans live (water and air pollution).

Not all of these elements depend on architecture, urban planning, and design, but a lot can be done by planning a human-friendly space for living, work, education, and recreation.

Urban lifestyle and built environment are considered unnatural. Therefore, it needs to be re-naturalized with nature-based solution to improve human health and well-being (Lawrence, 2021). Thus, quality of life is also related to contact with nature. The local microclimate and attractive public green spaces, located within walking distance, are important. It is necessary to ensure universal accessibility and proximity to services and the workplace. Mixed-use neighbourhoods can promote better life quality. However, it is worth noting that improving living conditions does not necessarily translate directly into an increase in life satisfaction. Life satisfaction is a subjective concept. Social capital, relationships, and spiritual needs (church within walking distance of the place of residence) are also important. The development of social capital can be supported by creating friendly public spaces, local cultural centres, and other places where people can stay in a pleasant, safe atmosphere. All these elements translate into human comfort, which in turn affects the level of attachment to the place of residence, as well as the property value and prices on the real estate market (Trojanowska, 2020).

There are two concepts of insecurities that seriously affect the quality of living: precariousness and precarity. Precariousness results from our interdependence and is manifested by universal insecurity. Precarity is the unequal distribution of precariousness (Boyko et al., 2021). The precarity is a failure of the social, economic, political, and spatial networks. The older generations are facing reductions of public services, casual labour contracts, and withdrawal of guaranteed pensions. The spatial distribution of precarity is concentrating in disadvantaged urban communities, where older people are facing multiple and interrelated issues. Experiences of precarity are negatively correlated with well-being. As the population is witnessing the increase in the total percentage of older people, the problems with precariousness and precarity are growing in importance. The spatial dimensions of precarity and precariousness additionally intersect with unequal distribution of material goods around the world. The Global Network of Age-Friendly Cities and Communities (GNAFCC), launched by the WHO in 2010, is seeking to prevent the exclusion of older generations (WHO, 2024).

Healthcare

The health indicators are benchmarks for sustainability of urban policies. They are linked to social equity, environment and development dimensions, accounting for house improvement, pollution (sanitation), transport (safety and sustainable modes), crime, equal access to healthcare services, fresh food and waste, as well as green spaces (Boyko et al., 2021).

Nicholas Freudenberg (2000) indicated three strategies that may improve the health of urban populations: increasing access and the quality of healthcare, reducing risk behaviour, and improving social conditions. Local healthcare should be able to treat non-life-threatening conditions, and reduce the burden on more specialized healthcare providers. The problem in most emergency departments (EDs) is that they receive substantial numbers of people with minor ailments, which impact on waiting times, and availability for more serious conditions (Boyko et al., 2021).

Access to basic healthcare services is one of the prerequisites of sustainable, mixed-use eco-neighbourhoods. The best would be to guarantee the location of basic healthcare services centrally, within 15-min walking distance (200–500 m) from every household.

Healthy housing

The notion of home refers not only to the concrete house, situated in the world, but also to the universe, encompassing a wide range of places whose meanings and scale can vary greatly, ranging from a house, a neighbourhood, a city, a region, a country, or the Earth (Serfaty-Garzon, 2023a). Houses and housing not only have the capacity to provide shelter, comfort, and respite, but also represent material culture, mentality, and relationship to the environment. Housing is built in line with cultural norms and socio-economic determinants (Serfaty-Garzon, 2023b).

Today, we are facing the growing mismatch between the offer of new housing and expectations of inhabitants. The interdisciplinary research evidence the need for diversity of households, housing units, domestic lifestyles, and housing preferences. At the same time the new housing construction is increasing in homogeneity to the point where it fails to address the fundamental nature of homemaking and living in the world (Lawrence, 2021). The 90% of the world's population is not served by the professional design community. They have little or no access to most of the products and services many people take for granted (Walker, 2008).

For housing, the location is everything. It implies social status (wealthy neighbourhoods, popular suburbs), well-being (density of healthcare, access to public parks), comfort and day-to-day life (local services, schools, sports and cultural facilities). The problem of environmental justice and heavy burden of pollution in disadvantaged neighbourhoods raise the question of a right to healthy environment (Serfaty-Garzon, 2023b).

The general housing design principles for good health and well-being are:

- Sufficient space, and layout to meet the needs of residents;
- Good environmental quality (air quality, hygrothermal, lighting, sound, state of repairs);
- Housing stability;
- Access to safe and good-quality facilities (green spaces, good schools, healthcare), and to employment opportunities;
- Opportunities to form meaningful connections with the local community (Boyko et al., 2021).

Housing quality is closely interrelated to improvements in health. Improving the quality of housing enhances children's health and attendance at school (Howden-Chapman et al., 2013). Housing has wide-ranging and long-lasting impacts on a child, including physical and mental health, development, and social well-being (Boyko et al., 2021). Poor housing qualities and parental stress, lack of safety, overcrowding, and inadequate facilities are responsible for a child's socio-emotional functioning, underachievement at school, and reduced physical activities. Living in persistent bad housing is negatively associated with each child's well-being outcome (Boyko et al., 2021). Overcrowding – too little space in relation to the number of residents – is impacting the health of an entire family. Besides the direct effects on physical health, and sickness, the mental health is also affected (sleep deprivation, lack of privacy, no place to study in quiet).

Bad housing may be associated with poor IAQ. There are several health issues related to poor IAQ: allergies, asthma, infectious diseases, cancer, and other genetic damage. IAQ is determined by the quality of the air outside the building; pollutants and emissions inside the building; the ventilation rate; the efficiency of filtration; and the maintenance of mechanical systems (Friedman, 2022). Poor IAQ may also be caused by volatile organic compounds (VOCs) emitted by building materials, finishing, furniture, equipment, or chemicals (paint, smoke, radon, asbestos, benzene). Not all VOCs have proven carcinogenic properties, but many of them cause other diseases such as allergies, asthma, mucous membrane irritations, headaches, and fatigue. Careful attention must be placed to choosing all indoor materials.

Lighting in buildings is of paramount importance in occupants' health and comfort. Both sunlight and artificial light are affecting human health, well-being, and productivity. The efficiency of a building in terms of its light illumination is most often determined using daylight factor (DF). DF is an indicator of daylight availability, which expresses the amount of daylight available as a percentage in the room (on the work surface), compared to the amount of unobstructed daylight available outdoors under cloudy conditions. A well-designed building, and clever use of natural lighting, can reduce electricity consumption by 30% (Bartosz, 2024). Optimizing the use of natural light in the interior spaces depends on many factors: the use of optimal glazing (glare control, insulation, permeability to solar radiation), the perception of variability of natural light, and visual comfort resulting from the quality of the view outdoors. Moreover, suitable, natural lighting contributes to the well-being. The location of windows and the size of glazing may contribute to both health promotion and energy savings. The location of largest glazing on the sunniest side (south on the northern hemisphere and north on the southern hemisphere) allows for fuller use of sunshine. Using smaller windows on the shaded side reduces energy losses (Zielonko-Jung & Marchwiński, 2012).

However, too much light and light pollution have several negative effects on health. It is impacting circadian rhythms, which regulate sleep and alertness (Boyko et al., 2021). The building should be designed and constructed in such a way as to reduce the risk of overheating during the summer. This requirement is considered to be met, if the windows, and other glazed and transparent partitions, are exposed to solar radiation in relation to geographical directions according to their location. There are several possible solutions for limiting excessive solar gains (blinds, curtains). Any device mounted on the exterior side, in front of the window is more efficient than all the interior solutions.

The thermal comfort is also significantly impacting the health and well-being. Evidence suggests that exposure to excessively low indoor temperatures can have negative impacts on physical health (cardiovascular and respiratory outcomes), mental and social health. Excessively high and excessively low-moisture indoors can also lead to adverse health effects (allergies, house dust mites, moulds, degradation of building materials). Maintaining comfortable temperature in rooms not only promotes human health and improves concentration, but also facilitates effective management of heat energy. Increasing the insulation of exterior partitions, using triple-glazed windows, and eliminating heat bridges improve the thermal comfort, while minimising energy losses. It is recommended to maintain the air temperature at an optimal level, but the principle "warm feet, cool head" used to promote underfloor heating is also justified by the physiology of the human body. Every heating system has its advantages and disadvantages, and must cover heating needs, but they differ the way warm air spreads (convection and radiation) and the surface temperature heating (Bartosz, 2024).

Effective ventilation must guarantee fresh air regardless of the conditions of weather and seasons and cover all rooms in the apartment. It is important to choose the right type of ventilation for specific needs. Ventilation should not only provide constant supply of fresh air, and removal of excessive moisture, without any drafts nor propagation of street noise, but also ensure that the heat (removed air) is not excessively wasted. Natural (gravitation) ventilation is the most widely used ventilation system. However, sometimes it is not sufficient to provide effective air flow and constant supply of fresh air. The mechanical ventilation with heat recuperation from exhaust air is promoted in passive building design (Zielonko-Jung & Marchwiński, 2012).

Acoustic quality is another basic element of health promotion, comfort, and well-being. From a user perspective, acoustic comfort is as important as thermal or IAQ. The quality of a room's acoustics depends on its form, size, and type of construction, as well as interior materials and finishes, number and type of equipment and furniture. Indoor acoustic is closely related with interior design. Exposures to loud noise can cause damage to hearing (reversible or permanent). Noise pollution has a particularly detrimental effect on children, both indoors and when playing outdoors. Numerous studies confirmed the negative effects of urban noise on stress-related health conditions and cardiovascular systems of children. Poor environmental quality, both indoors and outdoors, impacts children, as they have no control over their environment (Boyko et al., 2021).

Walking for life

Active living is one of the cheapest and most effective medicines. Walking, cycling, and exercising can have a positive effect on human body and mind. Today, one of the biggest health issues is the dual threat: obesity and malnutrition. Reduced urban mobility often goes in pair with increased consumption of energy-dense food and drink. Obesity is a health condition impacted by urbanization and malnutrition and is related to unhealthy diet and food deserts in disadvantaged neighbourhoods. Cities can promote both mobility and immobility that are key to well-being. The development of pedestrian paths in medium-density and mixed-use neighbourhood promotes active transportation and decreases car dependency. Local shopping supports local businesses and encourages sustainable economic development. Mobility is a significant contributor to people's experiences of well-being (Boyko et al., 2021).

Hope for everyone

The wealthiest population is declaring the highest level of well-being. Residents living in low-density, low-deprivation districts perceived their neighbourhoods to be aesthetically pleasing, accessible, and safe. People who are better off tend to have better access to resources and more control over their lives. They also had the most pride in their city and felt the least anxiety (Beaverstock et al., 2004).

The most vulnerable are people with poor health and low income. In majority of cases, they are living in high-energy-consumption housing without alternative options to upgrade their home, or leave. They should not be stuck at poor quality, low-energy-efficiency homes. Poor health is associated with higher energy usage at home, but reduced mobility and hence less energy for travelling. Any rises in energy costs would affect people with low income and poor health. They have larger home energy consumption because they lack alternative options (Boyko et al., 2021).

The design of liveable, healthy cities needs to provide affordable, spacious homes with good IAQ and access to greenspace.

The features of urban environment that have a positive impact on mental health are:

- green spaces (city parks and streets lined with trees)
- active spaces (walkable neighbourhoods and areas for recreation and leisure that promote both physical and mental health)
- social spaces that encourage people to gather (public squares with benches)
- safe spaces, where people have less concern about safety (crime, traffic) (Roe & McCay, 2021).

Well-being and happiness are all related to health, income, and education. Health and education have a prominent role to play, as they can stimulate economic growth. Therapeutic landscapes can promote mental and physical health, but political actions to provide a safe, liveable environment with decent education and satisfying income are also needed.

Note

1 World's Air Pollution: Real-time Air Quality Index, available online: https://waqi.info/, retrieved on 03.09.2024

References

Alexander Ch. 1977. *A Pattern Language. Towns – Buildings – Construction.* Oxford University Press, New York

Alexander, Ch. 2002. *The Nature of Order: The Process of Creating Life*. The Centre for Environmental Structure, Berkeley, CA.

Alexander, Ch. 2008. *Język wzorców*. Gdańskie Wydawnictwo Psychologiczne, Gdańsk, tłumaczenie z j.ang.

Balfour R., Allen J. 2014, September. *Improving Access to Green Spaces*. Health Equity Briefing 8. UCL Institute of Health Equity, Public Health England.

Bartosz D. 2024. *Podręcznik Audytora Certyfikatu ielony Dom v.2.01*. Polskie Stowarzyszenie Budownictwa Ekologicznego.

Beaverstock J., Hubbard P., Short J.R. 2004. Getting away with it? Exposing the geographies of the super-rich. *Geoforum*, 35(4), 401–407. https://doi.org/10.1016/j.geoforum.2004.03.001.

Bell S.L., Wheeler B.W., Phoenix C. Using geonarratives to explore the diverse temporalities of therapeutic landscapes: Perspectives from "green" and "blue" settings". *Annals of the American Association of Geographers*, 107(1), 93–108. https://doi.org/10.1080/24694452.2016.1218269.

Bickerstaff K., Walker G. 2001. Public understandings of air pollution: The 'localisation' of environmental risk. *Global Environmental Change*, 11, 133–145.

Bjerre L., et al. 2017. On-site and laboratory evaluations of soundscape quality in recreational urban spaces. *Noise & Health*, 19(89). Published by Wolters Kluwer, Medknow.

Boyko T., Cooper R., Dunn N. 2021. *Designing Future Cities for Wellbeing*. Routledge, London.

Charras K. 2023. Therapeutic environments. In: Marchand D. (Ed.), *100 Key Concepts in Environmental Psychology*. Routledge, Abington, Oxon, New York.

Cooper-Marcus C., Barnes M. 1995. *Gardens in Healthcare Facilities: Uses, Therapeutic Benefits, and Design Recommendations.* Univesity of California Press, Berkeley.

Cooper-Marcus C., Barnes M. 1999. *Healing Gardens, Therapheutic Benefits and Design Recommendations.* John Wiley and sons, New York. ISBN 0-471-19203-1.

Cooper-Marcus C., Sachs N. 2014. *Therapeutic Landscapes. An Evidence-Based Approach to Designing Healing Gardens and Restorative Outdoor Spaces*. John Wiley & Sons, Inc., Hoboken, NJ, pp. 14–35.

Corburn J. 2009 *Toward the Healthy City. People, Places, and the Politics of Urban Planning*. The MIT Press, Cambridge, MA, London, England.

Day R. 2006. Place and the experience of air quality. *Health & Place*, 13(2007), 249–260. https://doi.org/10.1016/j.healthplace.2006.01.002.

Dias de Oliveira J., Biondi D., Nunho dos Reis A.R., Viezzer J. 2021. Landscape visual and sound quality influence on noise pollution propagation in urban green areas. *Revista DYNA*, 88(219), 131–138. https://doi.org/10.15446/dyna.v88n218.94724.

Diehl L. 2017. Do All gardens heal the same? *CityGreen*, 14, 8–76.

Diehl L. 2023, Spring. A framework for categorizing healing gardens. *Culivate, Florida Horticulture for Health Network*, 3(2), available online: https://www.flhhn.com/uploads/1/3/8/6/138696150/spring_2023_cultivate_flhhn.pdf, retrieved on 12.09.2024.

Foo A.F. 1999. The seven lamps of sustainable city. In: Foo, A.F., Yuen B. (Eds.), *Sustainable Cities in the 21st Century*. Faculty of Architecture, Building & Real Estate, National University of Singapore, Singapore. pp. 109–130.

Freudenberg N. 2000. Time for a national agenda to improve the health of urban populations. *American Journal of Public Health*, 90(6), 837–40. https://doi.org/10.2105/ajph.90.6.837. PMID: 10846496; PMCID: PMC1446275.

Friedman, A. 2022. *Designing Innovative Sustainable Neighborhoods* (1st ed.). Routledge, London. https://doi.org/10.4324/9781003203025.

Gerlach-Springs N., Healy V. 2010, Spring. *The Therapeutic Garden: A Definition. Healthcare* and *Therapeutic Design Newsletter*. ASLA. American Society of Landscape Architects.

Gesler W. 1996. Lourdes: Healing in a place of pilgrimage. *Health & Place*, 2(2), 95–105.

Gesler W. 2005. Therapeutic landscapes: An evolving theme. *Health & Place*, 11, 295–297.

Gorman R. 2017. Smelling therapeutic landscapes: Embodied encounters within spaces of care farming. *Health & Place*, 47, 22–28. https://doi.org/10.1016/j.healthplace.2017.06.005.

Grylls T., von Reeuwijk M. 2022. How trees affect urban air quality: It depends on the source. *Atmospheric Environment*, 290, 119275. https://doi.org/10.1016/j.atmosenv.2022.119275.

Howden-Chapman F., Baker M., Bierre S. 2013. The houses children live in: Policies to improve housing quality. *Policy Quarterly*, 9(2), 35. https://doi.org/10.26686/pq.v9i2.4450.

James P. 2018. *The Biology of Urban Environments*. Oxford University Press, Oxford. https://doi.org/10.1093/oso/9780198827238.001.0001.

Jatzlauk G, Bartel S, Heine H, Schloter M, Krauss-Etschmann S. Influences of environmental bacteria and their metabolites on allergies, asthma, and host microbiota. *Allergy*, 72(12), 1859–1867. https://doi.org/10.1111/all.13220. Epub 2017 Jun 28. PMID: 28600901.

Karmanov D., Hamel R. 2008. Assessing the restorative potential of contemporary urban environment(s): Beyond the nature versus urban dichotomy. *Landscape and Urban Planning*, 86, 115–125.

Kahn P.H. 1999. *The Human Relationship with Nature. Development and Culture*, The MIT Press, Cambridge, MA, London, England.

Kavanagh J.S. 2011. A thumbnail history of therapeutic gardens in healthcare, LATIS 2011, Landscape Architecture Technical Information Series Number 1, Forum on Therapeutic Garden Design (2nd ed.), available online: https://www.asla.org/uploadedFiles/CMS/Practice/Research_Reports/ASLA_Research_TherapeuticGardenDesign_2011.pdf, retrieved on: 12.09.2024.

Kobylarczyk J. 2018. *Uwarunkowania środowiskowe w projektowaniu obszarów mieszkaniowych*. Wydawnictwo Politechniki Krakowskiej, Kraków.

Lawrence R.J. 2021. *Creating Built Environments: Bridging Knowledge and Practice Divides*. Routledge, Abingdon, Oxon, New York.

Lima M-L. 2023. Evaluating the landscape. In: Marchand D. (Ed.), *100 Key Concepts in Environmental Psychology*. Routledge, Abington, Oxon, New York. pp. 76

Lothian A. 2017. *The Science of Scenery: How We See Scenic Beauty, What It Is, Why We Love It, and How to Measure and Map*. CreateSpace Independent Publishing Platform, Scotts Valley, CA.

Louv R. 2005. *Last Child in the Woods: Saving Our Children from Nature-Deficit Disorder*. Algonquin Books, New York.

Lovasi G.S., Quinn J.W., Neckerman K.M., Perzanowski M.S., Rundle A. 2008. Children living in areas with more street trees have lower prevalence of asthma. *Journal of Epidemiology & Community Health*, 62(7), 647.

Maas J., Verheij R.A., de Vries S., Spreeuwenberg P., Schellevis F., Groenewegen P.P. 2006. Green space, urbanity, and health: How strong is the relation? *Journal of Epidemiology and Community Health*, 60(7), 587–592. https://doi.org/10.1136/jech.2005.043125.749.

Marsh E., et al. 2022. Endogenous mood state and hedonic responses to pleasant odors. *Journal of Sensory Studies*, 38, e12826. https://doi.org/10.1111/joss.12826.

McPherson E.G., Nowak D.J., Rowntree R.A. 1994. Chicago's Urban Forest Ecosystem: Results of the Chicago Urban Forest Climate Project. USDA Forest Service General Technical Report NE-186.

Millennium Ecosystem Assessment. 2003. *Ecosystems and Human Well-being: A Framework for Assessment*. Island Press, Washington, DC, available online: https://www.millenniumassessment.org/documents/document.356.aspx.pdf, retrieved on 28.08.2024.

Milligan C., Gatrell A., Bingley A. 2004. 'Cultivating health': Therapeutic landscapes and older people in northern England. *Social Science & Medicine*, 58, 1781–1793zz

Musy M. (Ed.). 2014. *Une ville vert. Les rôles du végétal en ville*. Éditions Quæ, Versaille.

Navarro O. 2023. Connection with nature. In: Marchand D. (Ed.), *100 Key Concepts in Environmental Psychology*. Routledge, Abington, Oxon, New York, pp. 25–26.

Nowak D.J., Crane D.E. 2000. The Urban Forest Effects (UFORE) model: Quantifying urban forest structure and functions. In: Hansen M., Burk T. (Eds.), *Integrated Tools for Natural Resources Inventories in the 21st Century. Proceedings of the IUFRO Conference*. USDA Forest Service General Technical Report NC-212, pp. 714–720.

Nowak D.J., Crane E.E., Stevens J.C., Ibarra M. 2002. *Brooklyn's Urban Forest*. USDA Forest Service General Technical Report NE-290.

Porteous, J. D. (1985). Smellscape. *Progress in Physical Geography: Earth and Environment*, *9*(3), 356–378. https://doi.org/10.1177/030913338500900303

Prior J. 2017. Sonic environmental aesthetics and landscape research. *Landscape Research*, 42(1), 6–17. https://doi.org/10.1080/01426397.2016.1243235.

Rayner G., Lang T. 2012. *Ecological Public Health. Reshaping the Conditions for Good Health*. Routledge, Abingdon and New York.

Roe J., McCay L. 2021. *Restorative Cities: Urban Design for Mental Health and Wellbeing*. Bloomsbury Publishing, London.

Sæbø A., Popek R., Nawrot B., Hanslin H.M., Gawronska H., Gawronski S.W. 2012. Plant species differences in particulate matter accumulation on leaf surfaces. *Science of the Total Environment*, 427–428, 347–354. ISSN 0048-9697. https://doi.org/10.1016/j.scitotenv.2012.03.084

Serfaty-Garzon P. 2023a. Home (chez-soi). In: Marchand D. (Ed.), *100 Key Concepts in Environmental Psychology*. Routledge, Abington, Oxon, New York. pp. 69

Serfaty-Garzon P. 2023b. Housing. In: Marchand D. (Ed.), *100 Key Concepts in Environmental Psychology*. Routledge, Abington, Oxon, New York.

Tallis M., Taylor G., Sinnett D., Freer-Smith P. 2011. Estimating the removal of atmospheric particulate pollution by the urban tree canopy of London, under current and future environments. *Landscape and Urban Planning*, 103(2), 129–138.

Trojanowska M., Sas-Bojarska A. 2018. *Health-affirming everyday landscapes in sustainable city. Theories and tools* Architecture Civil Engineering Environment Journal, ACEE2018, 11, 3, 41–52.

Trojanowska M. 2020. *Poszukiwanie standard projektowania ekoosieldi w Polsce*. Wydawnictwa Naukowe Uniwersytetu Technologiczno-Przyrodniczego w Bydgoszczy, Bydgoszcz.

Ulrich R.S. 1984. View through a window may influence recovery from surgery. *Science*, 224(4647), 420–421. https://doi.org/10.1126/science.6143402. PMID: 6143402.

Ulrich , R. S. (1999) Effects of gardens on health outcomes: theory and research. In: Cooper Marcus C. and Barnes M. (1999). Healing Gardens: Therapeutic Benefits and Design. Wiley (Series in Healthcare and Senior Living Design): New York, p. 624

Ulrich R. 2023. Stress reduction theory. In: Marchand D. (Ed.), *100 Key Concepts in Environmental Psychology*. Routledge, Abington, Oxon, New York, 143–461.

Velarde M.D., Fry G., Tveit M. 2007. Health effects of viewing landscapes – Landscape types in environmental psychology. *Urban Forestry & Urban Greening*, 6(4), 199–212.

Vos P., Maiheu B., Vankerkom J., Janssen S. 2013. Improving local air quality in cities: To tree or not to tree? *Environmental Pollution*, 183, 113–122. ISSN 0269-7491, https://doi.org/10.1016/j.envpol.2012.10.021.

Walker. 2008. Design for the other 90%, available online: https://walkerart.org/calendar/2008/design-for-the-other-90/, retrieved on 04.09.2024.

Ward Thompson C., Aspinall P., Bell S. 2010. *Innovative Approaches to Researching Landscape and Health. Open Space: People Space 2*. Routledge, Abingdon and New York.

WHO. 2024. About the global network for age-friendly cities and communities, available online: https://extranet.who.int/agefriendlyworld/who-network/, retrieved on 04.09.2024.

Williams A. 1998. Therapeutic landscapes in holistic medicine. *Social Science and Medicine,* 46(9), 1193–1203. https://doi.org/10.1016/S0277-9536(97)10048-X.

Williams A. 2010. Spiritual therapeutic landscapes and healing: A case study of St. Anne de Beaupre, Quebec, Canada. *Social Science & Medicine,* 70(10), 1633–1640. https://doi.org/10.1016/j.socscimed.2010.01.012.

Wilson Edward O. 1984. *Biophilia*. Harvard University Press, Cambridge, MA.

Wilson E. 2008. The nature of human nature. In: Kellert S., Heerwagen J., Mador M. (Eds.), *Biophilic Design: The Theory, Science, and Practice of Bringing Buildings to Life*. John Wiley & Sons, Inc., Hoboken, NJ, ISBN 978-0-470-16334-4.

Wilson J., et al. 2016. Urban park soundscapes: Association of noise and danger with perceived restoration. *Journal of Park and Recreation Administration*, 34(3), 16–35. https://doi.org/10.18666/JPRA-2016-V34-I3-6927.

Winterbottom D., Wagenfeld A. 2015. *Therapeutic Gardens. Design for Healing Spaces*. Timber Press, Portland, London.

WHO. World Health Organization. Regional Office for Europe. (1989). European Charter on Environment and Health, 1989. World Health Organization. Regional Office for Europe. https://iris.who.int/handle/10665/347390

Zachariasz A. 2006. *Zieleń jako współczesny czynnik miastotwórczy ze szczególnym uwzględnieniem parków publicznych*. Politechnika Krakowska, Kraków.

Zachariasz A. 2010. *Lyrical Beauty, Mistery and Peace – Historic and Modern Japanese Gardens*. Czasopismo Techniczne, 4-A, z. 12, r. 107. Wydawnictwo Politechniki Krakowskiej, Kraków.

Zielonko-Jung K., Marchwiński J. 2012. *Łączenie zaawansowanych I tradycyjnych technologii w architekturze proekologicznej*. Oficyna Wydawnicza Politechniki Warszawskiej, Warszawa.

Zimny H. 2005. *Ekologia miasta Agencja Reklamowo-Wydawnicza Arkadiusz Grzegorczyk*. Warszawa, ISBN 83-89961-30-X.

Zube E. 1987. Perceived land use patterns and landscape values. *Landscape Ecology*, 1(1), 37–45. https://doi.org/10.1007/BF02275264.

2 How to do it? Governance

How to do it?

Making cities health-promoting places is an evolving theme. The industrial revolution and post-World War II developments contributed to the depletion of natural resources and environmental degradation. A garden city model developed by Ebenezer Howard tried to give residents access to clean air, industry, and housing. To achieve this objective, they would be relegated to different areas of the city, creating suburbs far from central business districts and industries (Friedman, 2022). The suburban planning spurred urban sprawl, car dependency, consumerism lifestyle, and accelerated climate change. The environmental costs are accompanied by extensive social, physical, and psychological consequences. Long commutes to work, school, and essential services encourage a sedentary lifestyle with limited physical activity, which leads to chronic illnesses. Car-dependent urban planning resulted in inactivity, accompanied by food deserts, equals poor diet and obesity. The results are larger prevalence of cardiovascular diseases such as heart attacks and strokes – the leading cause of deaths worldwide (World Health Organization, 2021). Today, by learning from previous mistakes, future innovations can be developed and implemented.

Planning and management

Governance and management are the important pillars of sustainability. In urban planning, the governance is enforced with plans: regional, local, and master plans. In the last century, the planning was subject to major changes. The rules for planning were always based on improvement of living conditions for people. However, our knowledge about the natural resources and limits still needs a lot of development.

The most important improvement is to have objectives in planning. If promotion of human health, well-being, and happiness is the main aim of planners, it is a good starting point.

The planning process starts with survey to collect all of the relevant information about the city and region. It is necessary to make projections into the future as far as possible to predict the forthcoming needs and prospective directions of future development. The actual planning is the third stage. The plans are made taking into

DOI: 10.4324/9781003494898-3

account all the surveyed facts and interpretations to control future development according to principles of sound planning and the major objectives defined in first place (Hall & Tewdwr-Jones, 2020).

Regional planning

The objective of regional and urban planning is to specify a future state of regional and urban systems, in order to satisfy the human needs and prevent environmental, social, and economic damage. Plans need to be workable and feasible, to foster orderly and sustainable development. Regional planning is usually not related to producing blueprints for actual physical structures on the ground. Contemporary planning is focusing on processes, and not structures. Therefore, modern plans are models that represent the development from present into the predictable future. The difficulty of planning is born of conflicts of right against right. Planning means reconciling different set of values. Thus, planning is related to political decisions (Hall & Tewdwr-Jones, 2020). Nevertheless, Hall and Tewdwr-Jones (2020) recognize that political decision-making is a highly imperfect art. Jon Lang (2017) draws the attention that the major difference between planning in autocratic states and in democracies is that the decision-making in autocratic societies is beyond the control of citizens. Citizens in most nations have a right to vote for a national and local government, which is going to make planning decisions on their behalf. Participation in urban planning may be empowering to certain groups, if their voice is heard and they really could impact the spatial decisions included in the plan.

The usual flow of decisions is from regional planning to architectural design. The regional planning decisions are transferred into the city planning and translated into urban design. Jon Lang (2017) draws the attention to the fact that the flow of decision-making should not be unidirectional and should go in the other direction as well. The decisions at the detailed level impact the larger scale. Thus, they should move in an iterative manner from the neighbourhood level to the city level and then to regional level.

Urban planning

The most general city planning policy is represented in the comprehensive plan. It attempts to deal simultaneously with economic, social, environmental, and physical development (Lang, 2017). To implement comprehensive plans, the more detailed spatial law ordinances are developed: zoning codes, local plans, and master plans. The detailed spatial development codes specify the allowed and prohibited uses for a zone, building setbacks from the streets, the site coverage, percentage of biologically active area, floor area ratio (FAR) (the ratio of total usable floor area to the site size), the allowable height and size of buildings, the minimum number of parking places, colours and materials, roof type, and many other local conditions.

Jon Lang (2017) draws the attention to the fact that current zoning regulations throughout the world make it almost impossible to build urban tissue similar to Paris, London, and other well-loved cities. The pursuit of safe and salutogenic environments has some unintended effect on spatial segregation of functions and activities. Building a mixed-use, liveable, and vibrant city requires a modified approach that would concentrate on pedestrian needs, and not on transportation corridors for vehicles. It requires attracting mutually supportive uses.

All the prescriptions specified in planning ordinances must be specified with clarity, because they are used as legal basis for urban design and architectural projects.

Urban design

Jon Lang (2017) sees urban design as the mediator between planning and architecture, employed where planning meets architecture and landscape architecture. He describes four types of plan making and urban design work:

1 Total urban design – urban designer is part of the development team that carries the projects from schematic design to completion. The total urban design may range from new towns, new suburbs, new urban neighbourhoods to new housing developments. It is also used for campuses and historical revitalizations.
2 All-of-a-piece urban design – one team produces a master plan and sets the parameters and the developers work on components of the overall project.
3 Plug-in urban design – the infrastructure elements are designed first, and all the other developments plug into them.
4 Piece-by-piece incremental urban design – general policies and procedures are designed to steer the development in a specific direction.

Architectural design

Jon Lang (2017) warns that many new brilliant architectural designs do not relate to their surroundings and pay little heed to the public spaces and streets they are creating. Some of them seem to turn their back at the city depriving it of comfortable public spaces. He draws the attention to the fact that individual buildings, although they may significantly change the urban scene, are not urban design. Only complexes of buildings may be perceived as urban design. Jon Lang (2017) discerns four situations where individual buildings are regarded as urban design by architectural critics. The first is when buildings relate to their urban contexts. The second is when a building is a catalyst for urban development. The third is when it is a multi-use building. The fourth are large-scale architectural projects, with multiple buildings forming a complex (Lang, 2017).

In the city, in order for some buildings to become a foreground building, others which stay in the background are needed. The city of foreground buildings may become an unbearable cacophony. We need some foreground architecture as focal points, and dominants to organize our mental maps and facilitate orientation. Humans need certain level of stimulation, and we thrive in architectural variety. However, a sense of beauty, order, and organized form is needed. In order to manage the architectural variety, the instruments of planning are needed. Devising the design guidelines for urban design and architecture may promote well-functioning public spaces. They are directives that ensure that the objective of master plan or urban design scheme is put into practice.

Integrated design

Integrated design is a holistic approach to high-performance urban design, building design, and construction. It relies on collaborative work of every member of the project team representing different fields, who unite their skills from the very beginning of the conceptual phase to implement sustainability goals during the project (Harvard Energy & Facilities). The integrated design is the basis for the development of sustainable buildings. To achieve ambitious energy and environmental efficiency goals, the key is the selection of appropriately qualified specialists and the development of methods of close cooperation between them. Integrated design should involve a wide group of specialists, experts, and consultants from the very beginning (as opposed to traditional design, where the team expands as the design work progresses).

Design decisions are influenced by analyses and simulations, such as simplified energy modelling, analysis of daylight access and shading. In such a process, the investor and his consultants are active participants in the entire design process, during which they can constantly monitor the impact of decisions. All participants in this process have a real impact on the investment, and make decisions based on analyses, aware of consequences of implemented solutions. Previous experience shows that the costs of implementing sustainable buildings often turned out to be lower than planned, mainly due to appropriate exchange of information and joint decision-making at the earliest possible stage.

Eco-neighbourhood

Tomeu Vidal (2023) defines eco-districts as specific areas of a city, whose objective is to integrate the social, environmental, and economic dimensions of sustainability to accomplish the common sustainability objectives proposed by international organizations (Aalborg Charter, Leipzig Charter, UN 17 SDG). He mentions that eco-districts provide an opportunity to move sustainable experiences to a higher scale (city, region). The criticism of eco-districts is related to limiting the social dimension of sustainability, such as the inclusion of the marginalized and gentrification (Vidal, 2023).

Eco-neighbourhood is a way to operationalize the sustainable development. It is one of the methods to implement sustainable development principles in various countries and communities. There are various national and international classification systems and assessment tools to certify the sustainability of a neighbourhood. Sharifi and Murayama (2014) compared projects from Great Britain, the USA, and Japan and insist that there is no single universal standard to verify the sustainable development of a neighbourhood. Each of the systems can be accused of deficiencies in the assessment of social, economic, political, or legal aspects (Sharifi & Murayama, 2015). Geoff Boeing and his team (2014) noticed standard deficits in the assessment of many aspects of quality of life. Scientists believe that one standard of multi-criteria assessment is not sufficient as an indicator of the quality of a given housing estate.

The international certification standards for eco-neighbourhoods:

Building Research Establishment Environmental Assessment Method

The Building Research Establishment Environmental Assessment Method (BREEAM) system was created by BRE (Building Research Establishment) in Great Britain. It was launched in 1990 as the world's first multi-criteria environmental assessment system for new buildings. In 2011, the BREAM Communities certificate was developed for the assessment of social, environmental, and economic aspects of larger urban areas (residential and mixed-use districts). The certification scheme requires the implementation of sustainable solutions from the stage of initial schematic concept. Failure to comply with the BREAM Communities standards does not mean that the project would not receive a certificate. BREAM Communities standards allow for compensation by points in another category. The main problem with the BREAM Communities application is its complicated form, which makes it difficult to understand the nuances of the assessment.

Leadership in Energy and Environmental Design

The Leadership in Energy and Environmental Design (LEED) system was developed by the USGBC (U.S. Green Building Council) in the USA in 1993. The LEED certification system is used worldwide and includes an objective evaluation of results, carried out by a third party – the Green Building Certification Institute (GBCI). LEED for Neighbourhood Development (LEED ND) was implemented in 2009. The LEED ND system includes a base standard that must be met, and conditions, the fulfilment of which gives a certain number of points. LEED ND is criticized because it does not encourage local initiatives. The requirements included in the standard may cause the adopted solutions to be homogeneous, without reference to local conditions, which may result in the globalization of design choices in the spirit of New Urbanism. New Urbanism was created with projects in the USA in mind, and will not necessarily work in other local conditions. The system omits the possibility of working on improving processes. Today there are a few possibilities: LEED ND, LEED Communities, and LEED Cities.

DGNB

The German multi-criteria assessment system – DGNB Certification System – was created in 2008. It is based on LEED certification experience in the USA. The certification process requires the help of an accredited auditor. The residential neighbourhood assessment module is called DGNB Urban Districts. The assessment criteria can be divided into areas of sustainable development, including: environmental, economic, social aspects, and technical quality. Each of these areas accounts for 22.5% of the total point value. The quality of the process is 10%, and the localization is assessed separately, adding to final score.

La Haute Qualité Environnementale

The La Haute Qualité Environnementale (HQE) system (fr. La Haute Qualité Environnementale) was developed in 1996 in France. The assessment model for eco-neighbourhoods is HQE Aménagement™ and started in 2011. The system is constantly monitored and improved. The certification process is currently carried out by Certivéa, an auxiliary unit of the CSTB, Centre Scientifique et Technique du Bâtiment. Certivéa employs independent experts in the HQE certification process. To obtain the certificate, it is necessary to prove compliance with the requirements related to 17 points of the Sustainable Development Goals (SDGs). The HQE Aménagement™ concerns ecological, economic, and social areas.

The certification process can be carried out for both private and public operations. The following two conditions must be met:

- The project involves the construction of public utility facilities (school, swimming pool, public spaces, public roads, etc.).
- The project involves real estate transactions (purchase of land and/or division or consolidation of real estate).

The assessment is carried out by independent experts. It consists of conducting at least three audits: preliminary audit at the first stage, at least one audit during construction, and a final audit. Audits during construction phase are carried out annually, and in the case of shorter, smaller projects – every six and eight months.

In the HQE system, unlike LEED ND and BREAM, the aspect of continuous monitoring of the investment is emphasized. Post-construction assessments and user assessments, the so-called Retour d'Expériences, prepared by the Certivéa agency are particularly valuable. The reference of the HQE standard to ISO 37100 standards is clear. The HQE certificate is also the most universal in relation to the types of investments that can be subject to certification.

The HQE Aménagement™ and ÉcoQuartier certification standards are consistent with each other. Therefore, the 20 ÉcoQuartier requirements are compatible with the 17 HQE Aménagement™ points.

French national certification standard. ÉcoQuartier

The French government developed and implemented the national assessment system ÉcoQuartier in 2012. According to the MEDDTL French Ministry of Ecology, Sustainable Development, Transport and Housing, ÉcoQuartier is the development of a residential neighbourhood – a part of the city, within which sustainable solutions have been applied: transportation, functional diversity and various forms of development, and eco-construction, but at the same time it facilitates the co-habitation of different social groups and social participation.

The programme consists of four stages of granting the certificate:

1 the first stage is the signing of the eco-neighbourhood intentional letter (Charte ÉcoQuartier) by members of the board of the given commune and cooperating entities;
2 the second stage includes the preparation of the project, its assessment and verification of compliance with the programme requirements. The projects are assessed by independent experts. At this stage, it is still possible to make corrections during the construction phase;
3 the third stage is the awarding of the ÉcoQuartier certificate by the national commission at the request of the regional commission, after presenting the experts' assessment.
4 the fourth stage is the assessment of the functioning of the estate three years after receiving the certificate. It is checked how in practice the residents and managers of the area fulfil their obligations.

The ÉcoQuartier programme favors the exchange of experience between communes creating eco-neighbourhoods. It envisage the use of the help of experts.

The ÉcoQuartier programme is an attempt to operationalize the idea of sustainable development in French territory. It was developed by Franck Faucheux. After many studies, he came to the conclusion that the way to optimize the initial costs of developing eco-neighbourhood was to use the scale effect, and purchasing materials in bulk for the entire district, and not for a single building. Many solutions can be balanced on a district scale, which leads to cost reduction. Promoting the pedestrian-oriented design means there would be no paved internal roads and only stabilized dirt paths for occasional passage of emergency vehicles. In this case, savings concern not only building materials and road construction costs, but also the costs of rainwater management, which are minimized when natural infiltration in the ground is possible. The installation of solar collectors and photovoltaic panels can be more economical in the case of a larger number of recipients. It is therefore a new type of space management, which allows for sustainable pro-ecological management on a district scale in relation to the technical solutions used in individual buildings. Frank Faucheux saw the potential of this strategy. He knew that proposing a certification strategy at the national level could allow for its dissemination and would encourage its use on a district and neighbourhood scale. In this way, it was possible to facilitate the introduction of pro-ecological initiatives in building designs and to popularize their use on a larger scale (Trojanowska, 2020).

The ÉcoQuartier system was created to enable a wide range of local government entities to create sustainable neighbourhoods, and this goal has been achieved. It can serve as a model solution.

There has also been criticism of the French ÉcoQuartier certification system. It concerns, similarly to other multi-criteria assessments (LEED, BREAM, HQE), the selectivity and subjectivity of the assessment. A neighbourhood can meet many of the assessment criteria and receive a certificate, and yet still be very burdensome for the environment in other aspects. The French research community mainly criticizes the lack of public participation in the design of eco-neighbourhoods. Future residents receive a finished product, they do not participate in the integrated design process. Another problem was the threat of unification and standardization, while architecture and urban planning are fields in which each project should be a prototype and proof of progress in the search for the best possible solutions.

The French state is a welfare state, and social benefits are very high. In France, the state administration is centralized, strong, and well-developed. It is the state that dictates the conditions, not the developer. Eco-neighbourhoods are built on public land and initiated by local government administration. There is a regulation on social solidarity – 20% of apartments in a commune must be designated as social housing. Therefore, French eco-neighbourhoods usually have a large share of public housing built with public money. On the other hand, the architectural education of society, which has been conducted for years, has brought good results. French citizens are aware of their rights in terms of protecting spatial order (Trojanowska, 2020).

The certification process in the ÉcoQuartier standard is a good illustration of the systemic approach. The drive is for continuous improvement of processes. That is why the assessment of sustainable solutions is never closed, but constantly monitored and re-evaluated (Trojanowska, 2020).

Inspiring policies

The term sustainability was coined in the mid-20th century with raising concerns about the diminishing capacity of the Earth. It was developed through a series of United Nations (UN) meetings. A special commission to study global sustainability was established – The World Commission on Environment and Development (WCED), later renamed the Brundtland Commission. In 1987, the report Our Common Future was published with the definition of sustainable development: “development that meets present needs without compromising the ability of future generations to meet their own needs” (Brundtland, 1987, p. 41). The five key pillars of sustainability are social, economic, environmental, cultural, and governance (Brundtland, 1987).

In 1978, the WHO Healthy Cities initiative, inspired by WHO European Health for All strategy and the Health21 targets, was developed (Friedman, 2022). The long-term goal of the WHO Healthy Cities project is to integrate health in the agenda of policy decision-makers in cities. The project strives to create a strong partnership for health promotion and to apply a community-based participatory approach (Lawrence, 2020).

WHO’s Age-Friendly Cities and Communities was formalized in 2006 (Boyko et al., 2021; WHO 2024a, 2024b). The mission of these initiatives is to put public health on the political and social agenda of cities around the world. The goal is to improve the relation between people and their needs, and the environments in which they live, work, and recreate, to foster health.

In 2010, the WHO published Global Recommendations on Physical Activity for Health (WHO, 2010). This document recognizes physical inactivity as the fourth leading risk factor for global mortality. The rising levels of physical inactivity have implications for the prevalence of noncommunicable diseases (NCDs). This report aims to provide guidance on the frequency, duration, intensity, type, and total amount of physical activity needed for the prevention of NCDs. The primary target audience for these recommendations are policymakers at national level.

The UN 2030 Agenda for Sustainable Development objective is to transform economic and social processes that damage or destroy natural and life-support systems for all biological organisms living on planet Earth by addressing 17 SDGs and 169 targets (Lawrence, 2020). Particularly one of the SDGs, proposed by the UN: 11. "Make cities and human settlements inclusive, safe, resilient and sustainable" and Point 11.7 "By 2030, provide universal access to safe, inclusive and accessible, green and public spaces, in particular for women and children, older persons and persons with disabilities" resonates with the concept of therapeutic landscapes. Numerous studies confirm the health-promoting qualities of contact with nature and problems resulting from the deprivation of access to public green spaces (Maas et al., 2006, 2008, 2009; WHO, 2016). Easy access to safe and inclusive public green spaces is still one of the long-lasting problems of urbanized areas around the globe. The major question is how to implement this goal in practice and design cities to provide easy access to safe and inclusive public green spaces. The standard for health-promoting urban places presented in Chapter 1 was developed to help designers, planners, and policymakers to improve the quality of public spaces.

The building construction sector can make a fundamental contribution to sustainable development. It relies heavily on natural resources, substantial energy consumption, and produces important amount of waste. Limiting the consumption and increasing the recycling rate would make a huge impact on the world scale.

In Europe, one of the major goals is the promotion of energy-efficient construction and improvement of the energy performance of buildings. It was proclaimed in the first Directive 2003/91/EC of the European Parliament and of the Council EPBD (Energy Performance Buildings Directive) was issued on December 16, 2002.

The second EPBD, Directive 2010/31/EU, contains the definition of a nearly zero-energy building (NZEB). According to the directive, it is a building with very high-energy performance, in which a significant part of the energy should come from renewable sources. Determining what does it mean almost zero is left to the individual Member States. The NZEB building is an investment aiming to minimize the total cost (investment costs + costs related to the operation of the building and its technical systems).

The new, third EPBD Directive 2018/844 of the European Parliament and of the Council of May 30, 2018, amended the previous two. The directive includes:

- a long-term strategy for the renovation of existing housing resources in order to achieve high energy efficiency corresponding to the standard of buildings with almost zero energy consumption by 2050 (the so-called NZEB standard according to EPBD 2010/31/EU);

- electromobility (promotion of the use of electric vehicles);
- technical building systems, which specify the definitions of building automation and control systems using energy from renewable sources; and
- new methodology for determining energy performance.

On April 12, 2024 The Council of Europe finally adopted the changes in the EPBD to significantly accelerate the decarbonization of the construction industry. This is one of the key legislative elements of the European Green Deal and the Fit for 55 package. The objective is to eradicate energy poverty and reduce greenhouse gas emissions. In accordance with new provisions, from 2030 all new buildings will have to be zero-emission. The zero-emission building is characterized by very low energy consumption, which is entirely covered by renewable energy sources (RES). To comply with this standard, renewable energy systems in buildings must be installed. Therefore, there is an increase in popularity of heat pumps, solar collectors, photovoltaic installations, and small wind turbines.

European Charter on Environment and Health (1989) was adopted to promote the environment conducive to the highest attainable level of health and well-being, recognizing that this is the entitlement of each individual (WHO, 2024b). This charter also guarantees access to information on the state of the environment and the participation in the decision-making process.

The term "net zero" is commonly used to describe energy-efficient dwellings that produce as much energy as they use (Friedman, 2022). The net-zero neighbourhood would produce as much energy as it uses. The most effective is the centralized energy infrastructure for heating and cooling. It allows optimizing the use of fossil fuels and RESs and making advantages of the economy of scale in pursuit of the most efficient innovative solutions. The residual energy may be stored in the energy store or returned to the central facility to be redistributed later. Many of the solutions related to design are more easily managed at the community scale. The individual zero-energy houses are highly costly, and majority of people would not be able to afford living in them. Net-zero neighbourhoods are a more economical option, thanks to optimalization related to economy of scale and sharing resources (Friedman, 2022).

It is worthy to reduce environmental (ecological) footprints: the carbon footprint and the water footprint of the planned investment. Environmental footprints compare human consumption with the Earth's ability to regenerate. The environmental footprint is the estimated number of hectares of land and sea surface needed to compensate for the resources used to consume and absorb waste. It is measured in global hectares (gha) per person. To measure an ecological footprint, the calculation considers every extraction from nature: the water and grain required to feed livestock and farm fish; the land required to grow food; extraction of resources such as mining and timber; and so on. To promote sustainability, reasonable extraction of resources is permitted if there is sufficient protected land to balance emissions and support biodiversity. However, ecological reserve, with more natural land, must be protected and preserved (Friedman, 2022). The problem with urban areas is that they occupy large surfaces of land, and their ecological footprint is even larger. The area needed to sustain them and assimilate the waste they produce exceeds their surface many times (Lawrence, 2020).

The carbon footprint is the total greenhouse gas emission by the actions and consumption of a person or larger entity, such as a company or community, or over the full life cycle of an investment. Water footprint is the total freshwater consumption of a person or larger entity. It is a multidimensional indicator and can be conventionally divided into blue, green, and grey. The blue water footprint is the consumption of freshwater from rivers, lakes, ponds, and underground reservoirs. The green water footprint is defined as that part of the volume of rainwater that was taken up by plants to produce specific products (agriculture). The grey water footprint refers to water pollution and is calculated as the volume of water that would be needed to dilute the discharged pollutant load to such an extent that the quality of the resulting water does not exceed established standards.

The life-cycle cost analysis (LCCA) is a helpful approach which takes into consideration the destruction of natural environment. LCCA analysis makes it possible to estimate the costs generated throughout the entire life cycle of products: costs of obtaining raw materials (extraction, processing, and transport), production of building materials, their incorporation into the building, use (necessary repairs) and demolition, recycling, and waste disposal. Energy supply and consumption as well as transport costs are also calculated. The analysis takes into account the costs of purchasing, owning, and selling a building or structure. LCCA analysis makes it possible to estimate, at an initial stage, various design variants in terms of the costs of individual stages of the investment's life. It is also valuable to compare the real costs of using modern materials with the costs of traditional solutions and construction solutions in terms of savings throughout the entire life cycle, especially at the operational stage. This allows comparing different project variants that may differ in initial or operational costs.

The main difference between traditional billing of investment and the LCCA life-cycle cost calculation is expanding the perspective of cost analyses, and taking into account not only the investment with operating costs, but all costs, including the demolition of a building or structure.

Public participation

It is very important to develop an attitude of participation among residents in the process of creating an eco-neighbourhood. The idea is to make them feel responsible for the success of the investment. A key factor in striving for sustainable development is people's responsible attitude towards the ecology of everyday life. If the question is when should the public participation be engaged in the planning process, the answer would be as early as possible. The earlier the people are engaged in the process, the less opposition it would have once it is finally developed.

The public opinion is critical to initiate change in a community. Sustainable development involves a change in the approach to city planning. The traditional role of the state, which plans at the central level, manages projects and finances investments, is changing to a new attitude of the accompanying state, which supports grassroots local initiatives and accompanies the sustainable development of the urban fabric. Pope Francis expressed his views on the pursuit of sustainable development in his encyclical Laudato Si (2015). He expressed it in the words: caring for our common home. This is the kind of caring, stewardship approach that is needed. Eco-neighbourhoods provide people with the opportunity to "think globally and act locally" and actively take responsibility for the future of our planet. This leads to a sense of freedom and the ability to decide about the future. Pope John Paul II wrote that people are becoming more and more aware of the limited amount of natural resources and the need to respect the laws of nature when planning further economic progress. Economic prosperity, if not guided by moral and ethical principles, quickly turns into human enslavement (Pope John Paul II, 1987).

Action is needed at the local level. The sense of responsibility is associated with satisfaction with the place of residence, trust in local authorities, and belief in one's own abilities. Public participation promotes the transparency of decisions and public finances. Cooperation with residents and users is very important to create an eco-neighbourhood. You cannot expect residents to design their neighbourhood themselves, but they must be allowed to express their needs. Even though the professionals may have the technical knowledge, the local populations have the intricate understanding of their environment. It is crucial to collect and implement that knowledge. Meeting residents' expectations often does not entail additional costs, but only allows them to choose the best from a range of equivalent solutions. Residents are most likely to choose proven and widely known solutions. It is difficult to achieve social acceptance of an experimental solution before the prototype is launched. However, you can try to obtain social consent to implement the innovation after presenting the arguments and results of pilot studies. In the case of eco-neighbourhoods, it is assumed that residents will be informed about the progress of design work, and would participate in information meetings, workshops, and submit their comments to the project. Public participation should encompass the entire span of the project, from initiation till completion, use, and management.

Jason Corburn (2009) explains that the public participation process offers planners numerous opportunities to promote social cohesion and build social networks, especially between different groups (social capital bridging). This possibility is often overlooked. Jason Corburn also writes that healthy places are doubly constructed; physically (the "built environment": buildings, streets, parks, etc.) and socially (attribution of meanings, interpretations and narratives, construction of networks, and processes to shape these meanings and outcomes).

Therefore, health promotion requires "placemaking" – strengthening the relationships between people and the places where they live.

Public consultation should be reflected in the decision-making process. Participation means education about what is realistic and feasible. Once the realistic expectations of the public are expressed, they should be meet. It is a very empowering experience to have real impact on surroundings through the participation process. Public participation can have a very beneficial effect on mental health and physical residents, if their demands are met and they have a real impact on the final effects. Working for the well-being of your own community feeling of belonging can have a very positive impact on the sense of agency and creating stronger social bonds (Roe & McCay, 2021). It also contributes to create a sense of self-confidence. Transparency of procedures and ensuring real influence on the course of events are crucial. On the other hand, it may have a negative health effect, if people are actively participating only to find out that their postulates were not taken into consideration. If it turns out that the participating residents were only consultants, whose voice is not taken into account, this may have a negative impact on their well-being (Roe & McCay, 2021).

Processes

In designing eco-neighbourhoods, the key issue is to move away from the primacy of designing spatial structures to designing processes. Both the general plan (master plan) and the architectural solutions for neighbourhoods and individual buildings must take into account the possibility of future changes in the design. Unpredictable situations, technological progress, climate change, etc. stimulate the evolution of the urban and architectural fabric. It is necessary to provide for the flexibility of development forms, and methods of use to enable easy adaptation to the changing needs of users. Management of processes is one of the pillars of sustainable development. In the case of eco-neighbourhoods, a new approach is to treat them as ecosystems and analyse their entire life cycle and total cost.

Designing eco-neighbourhoods creates the opportunity to introduce innovative ecological, social, and economic solutions. Innovations also concern the way of shaping the urban fabric and the architectural and urban form. The implementation of innovative solutions, especially in cooperation with the academic community, should be rewarded in the evaluation process. Introducing innovations should be economically justified. The best solution is the most cost-effective in implementation and maintenance, taking into account the full environmental costs (LCCA). Rational use of resources requires considering various solutions in terms of the life costs of buildings and the entire investment. Assessment of different variants by all stakeholders, including representatives of the social community, can help in choosing the optimal solution.

Caring for sustainable development means long-range planning and taking into account changing needs over time. All architectural solutions should be durable and have the potential to change without incurring additional costs. Preparing for future changes involves designing adaptive structures that can be easily adjusted to new tasks. You can check the variants of the adopted solutions (combining and dividing apartments to meet the needs of a changing number of residents, freedom to divide service spaces). Modular structures, mobile walls, and a light frame in interiors are solutions that allow for adaptation. Multifunctional structures can be more easily adapted to the changing needs of users over time. It is worth asking whether apartments can be combined for the needs of large families and then divided when the children move out of the house, and the parents no longer need more space, or they want them to run their own household in a separate apartment? Will it be possible to convert the office building into a hotel, school, or hospital?

It is worth answering these questions at the design stage. A very fashionable word nowadays is resilience, i.e., the ability to overcome difficult moments. In the case of eco-neighbourhoods, it is worth asking what is the greatest potential risk that can be predicted and can you prepare for it?

It is helpful to establish a programme and goals related to sustainable development and health promotion at the very beginning of discussions about the planned investment. It is crucial to implement an action strategy that is transparent and feasible. It is important that the adopted management strategies for the eco-neighbourhoods project are transparent to all stakeholders.

In order to become an effective instrument, the eco-neighbourhood must be recognizable and highly valued in society. This is a sign of the quality, which must inspire trust. Appointment of independent experts, visits to the construction site, talks with stakeholders, transparency of information for residents and the public, etc. are all important for eco-neighbourhood acceptance.

It is worth considering what can be done to improve social awareness. Activities at the national level are important: promotional and educational campaigns, educational activities in the mass media, elements woven into teaching curricula in public schools, etc. are all important.

Participating in the creation of an eco-neighbourhood can be an opportunity for training and development for all stakeholders. Open meetings of all interested parties, as well as thematic and industry discussions, are an opportunity to meet, and develop common strategies, but also to learn. It is worth inviting experts and reserving time for lectures at each open meeting. The example of eco-neighbourhoods in France shows that organizing regular meetings (every week) brings good results. Cyclical meetings turned out to be very popular, residents eagerly come to listen stakeholders' discussions, and they join the discussion to express their opinions. The meeting places are school facilities, local cultural centres, or, for example, post-industrial facilities adapted for permanent or temporary purposes. The need to organize meetings regularly, starting from the initial phase of the planned investment, should be considered a priority that may affect the work schedule, e.g., the first building put into use in the eco-neighbourhood may be a cultural centre with a multimedia room adapted for meetings to enable organizing them at the investment site as quickly as possible.

A good way to look for better solutions is education and mutual exchange of experiences. In France, the ÉcoQuartier club was organized, which became a platform for exchanging information between neighbourhoods participating in the certification. The eco-neighbourhood project would aim to educate wide circles of society and improve the pursuit of sustainable development throughout the city, not only the investment area (Trojanowska, 2020, 2023).

References

Boeing G., et al. 2014. LEED-ND and livability revisited. *Berkeley Planning Journal*, 27, pp. 31–55.

Boyko T., Cooper R., Dunn N. 2021. *Designing Future Cities for Wellbeing*. Routledge, London.

Brundtland G.H. 1987. Our Common Future: Report of the World Commission on Environment and Development. Geneva, UN-Dokument A/42/427, available online: https://digitallibrary.un.org/record/139811?v=pdf, retrieved on 05.09.2024.

Corburn J. 2009. *Toward the Healthy City. People, Places, and the Politics of Urban Planning*. The MIT Press, Cambridge, MA, London.

European Charter on Environment and Health. 1989. Available online: https://www.health.belgium.be/sites/default/files/uploads/fields/fpshealth_theme_file/8562404/European%20Charter%20on%20Environment%20and%20Health.pdf, retrieved on 17.09.2024.

Francis, The Holy Father. 2015. Encyclical Letter *Laudato Si'* of the Holy Father Francis on Care for our Common Home, available online: https://www.vatican.va/content/francesco/en/encyclicals/documents/papa-francesco_20150524_enciclica-laudato-si.html, retrieved on 11.9.2024.

Friedman, A. 2022. *Designing Innovative Sustainable Neighborhoods* (1st ed.). Routledge, London. https://doi.org/10.4324/9781003203025.

Hall P., Tewdwr-Jones M. 2020. *Urban and Regional Planning*. Routledge, Abingdon, Oxon, New York.

John Paul II, Pope. 1987. Encyklika Sollicitudo Rei Socialis. Watykan (SRS 27–28), available online: https://adonai.pl/jp2/pliki/srs.pdf, retrieved on 10.06.2024.

Lang J. 2017. *Urban Design*. Routledge, Abingdon, Oxon, New York.

Lawrence R.J. 2020. *Creating Built Environments: Bridging Knowledge and Practice Divides*. Routledge, Abingdon, Oxon, New York.

Maas J., Verheij R.A., de Vries S., Spreeuwenberg P., Schellevis F., Groenewegen P.P. 2006. Green space, urbanity, and health: How strong is the relation? *Journal of Epidemiology and Community Health*, 60(7), 587–592. https://jech.bmj.com/content/jech/60/7/587.full.pdf

Maas J. 2008. *Vitamin G: Green Environments – Healthy Environments*. Nivel, Ultrecht.

Mass J., van Dillen S.M.E., Verheij R.A., Groenewegen R.P. 2009. Social contacts as possible mechanism behind the relation between Green space and health. *Health and Place*, 15(2), 586–595.

Maas J., Verheij R.A. 2009. Morbidity is related to a Green living environment. *Journal of Epidemiology and Community Health*, 63, 967–973.

Roe J., McCay L. 2021. *Restorative Cities: Urban Design for Mental Health and Wellbeing*. Bloomsbury Publishing, London.

Sharifi A., Murayama A. 2014. Neighborhood sustainability assessment in action: Cross-evaluation of three assessment systems and their cases from the US, the UK, and Japan. *Building and Environment*, 72, 243–258. https://doi.org/10.1016/j.buildenv.2013.11.006.

Sharifi A., Murayama A. 2015. Viability of using global standards for neighbourhood sustainability assessment: Insights from a comparative case study. *Journal of Environmental Planning and Management*, 58(1), 1–23. https://doi.org/10.1080/09640568.2013.866077.

Trojanowska M. 2020. *Poszukiwanie standard projektowania ekoosieldi w Polsce*. Wydawnictwa Naukowe Uniwersytetu Technologiczno-Przyrodniczego w Bydgoszczy, Bydgoszcz.

Trojanowska M. 2023. *Projektowanie zielonych przestrzeni publicznych*. Wydawnictwo Naukowe PWN, Warszawa.

Vidal T. 2023. Eco-district. In: Marchand D. (Ed.), *100 Key Concepts in Environmental Psychology*. Routledge, Abington, Oxon, New York, pp.151.

WHO. 2010. Global recommendations on physical activity for health, available online: https://www.who.int/publications/i/item/9789241599979, retrieved on 07.09.2024.

WHO Regional Office for Europe. 2016. Urban green spaces and health, available online: https://iris.who.int/bitstream/handle/10665/345751/WHO-EURO-2016-3352-43111-60341-eng.pdf?sequence=3&isAllowed=y, retrieved on 13.08.2024.

WHO. 2021, June 11. Cardiovascular diseases (CVDs), available online: https://www.who.int/news-room/fact-sheets/detail/cardiovascular-diseases-(cvds), retrieved on 23.07.2024.

WHO. 2024a. About the global network for age-friendly cities and communities, available online: https://extranet.who.int/agefriendlyworld/who-network/ /, retrieved on 04.09.2024.

WHO. 2024b. Healthy cities, available online: https://www.who.int/southeastasia/activities/healthy-cities, retrieved on 05.09.2024.

3 Ecology

"Act locally think globally" is important for eco-neighbourhoods. Local actions for the global future rely on planning different sizes of urban parks – large-scale, district parks, and pocket gardens. Human health and well-being are influenced by local microclimate. Public health promotion is related to everyday, effortless contacts with nature for each person. Humans are part of natural ecosystems. The actions that promote the sustainability of local environment are also good for human health. The issues of health promotion would vary around the globe. There is still a long way to go to offer everyday health-promoting places, but the journey has already started. The changes on a global scale are initiated locally.

Blue and green infrastructure

The grid of blue and green infrastructure is a backbone of every urban development. It is formed with urban green spaces, such as public parks, playgrounds, and private greenery, and a grid of green corridors – green streets (tree-lined alleys), green walls, and green roofs. The most important for well-functioning of green and blue infrastructure is the continuity of corridors. Corridors should be interconnected with each other and with the urban green spaces. The public open green spaces (POS) should be organized in a hierarchical manner – large-scale urban parks, mid-size district parks, and small pocket parks or squares. The residential greenery and private gardens are complementing the system. As a general rule, the best for the biodiversity promotion and human health are the open green spaces which are larger in dimensions, thus have enough space for development and maintenance of a self-sustainable population. It is worth recalling that the minimum size of a sustainable population, i.e., the smallest number of individuals that provides a safe level of probability survival in a given period of time is 1,000, so it can be estimated that this is the minimum number of tree seedlings for re-naturalization of urban forest. A shape similar to a circle is better than narrow strips because it allows for the reconstruction of the habitat and the development of natural ecosystems (Trojanowska, 2020).

The inter-connectivity is important for the promotion of biodiversity and exchange of genetic material. However, there is one restriction though. The

DOI: 10.4324/9781003494898-4

vulnerable ecosystems that were developed in seclusion should stay that way. Opening them to form part of the infrastructure may destroy them. That ecosystem may not be resilient enough to new pathogens, coming with human penetration or domestic animal presence. Therefore, enclosing some parts of the green infrastructure with fences to prevent the trespassing by humans or domestic animals is envisaged in some situations.

The health and social benefits of public green spaces offer a value that is incommensurable with the monetary value (Louv, 2005; Zhang et al., 2017). The influence is on stimulating social contacts and cohesion, lowering the level of crime, and preventing social exclusion (Lawrence, 2020; Maas et al., 2009a, 2009b). There are also interesting studies on the correlation between the amount of greenery and reduction in incidents of antisocial behaviour, i.e., violence and vandalism. It has been proven that the greener the surroundings of a building, the lower the number of crimes committed there, i.e., theft, vandalism, aggression, etc. (Garvin et al., 2012; Kuo, 2010; Kuo & Sullivan, 2001; Kuo et al., 1998). Therefore, the operational costs of landscaping and maintenance should be measured using a life-cycle assessment of both financial and non-monetary costs and profits.

The provision, size, and shape of green and blue infrastructure are the outcomes of political choices (Lawrence, 2020). These choices are the foundations of public policies, plans, documents, and law regulations. The coefficient ratio of the area of biologically active green areas to the entire area of land is important. It may be regulated and stipulated by zoning and local development plans. In areas of dense urban development, some concessions are being made. In some circumstances, the green roofs and green walls count as a fraction of biologically active green areas on the ground.

The quality of green and blue infrastructure, and the possibility of subsistence of an ecosystem, depends on many factors: soil condition, topography, wind and sun paths, existing flora and fauna native to that area (Friedman, 2022). Retaining existing vegetation can support the unique characteristics of the ecosystem and strengthen its resilience. Enhancing and nurturing ecosystem services is not only ecologically and socially fair, but also economically viable in the long term (Lawrence, 2020).

Accessibility of public green spaces

The objective to "provide universal access to safe, inclusive and accessible, green and public spaces, in particular for women and children, older persons and persons with disabilities" is one of the targets (SDG 11.7) of the 2030 Agenda for Sustainable Development. There are several steps that must be taken to guarantee the fulfilment of that objective. The layout and landscaping of urban environment for the physical activity of children, youth, adults, and older people have certain requirements, e.g., safe pedestrian streets, cycle lanes, playgrounds, and public parks with places for all age groups to be active.

According to Frances Kuo (2010), the question of contemporary urban planning and landscape architecture is not "whether a resident can have regular access to green spaces", but "whether a resident actually has regular access to them". Contact with green environment, in order to have a positive effect on health, must be ensured every day. Frances Kuo recommends creating as many contacts with nature as possible, bringing nature closer to people and bringing people closer to nature (Kuo, 2010). Jolanda Maas (2008) developed the concept of vitamin "G" – green – which should be taken daily. Hartig et al. (2014) emphasize that access to green spaces reduces stress and improves mental health of city residents. Studies have shown that access to green spaces can reduce the incidence of cardiovascular diseases and improve mental health (Maas et al., 2006; Mitchel et al., 2008).

Thus, access to safe, good-quality playgrounds and public parks is especially important for children who are educated in the era of virtual reality and gaming technologies. The perception of reality and environment needs to be grounded in more traditional down-to-earth outdoor physical activities (Boyko et al., 2021). Charles Montgomery in his book *Happy City* (...) (2015) states that if there is a park less than 800 m from the house, the probability that school children will go there is high. If the distance is longer, children wait for their parents to pick them up.

Hierarchy of urban parks – from large scale to pocket parks

We need large-scale parks, neighbourhood parks, and pocket parks ranging in size to ensure that urban dwellers have access to spaces that meet their specific needs. However, we have to keep in mind that even several small pocket parks, each with an area of less than 1 ha, cannot meet the recreational needs of residents to the same extent as one larger district park (Trojanowska, 2017, 2023a, 2023b). It is worth to remind that even ten pocket or neighbourhood parks will not replace one large scale park with an area significantly exceeding 10 ha. This does not mean that it is not worth creating pocket parks. Everyone should have guaranteed access to the pocket park within 7 min walking distance (200 m), district park within walking distance of 15 min on foot (450 m), or other available means of transport, and a large scale park area within 15-30 min on foot, or other accessible means of transport. The distance can be increased by improving the public transport, and sometimes also the walkability, quality, and connectivity of pedestrian and bicycle paths. The pocket parks complement the functions of large scale parks, when they are located in the immediate vicinity of users. They are important for everyday recreation. Easily accessible parks help to prevent the detrimental psychological effects of isolation, long commutes, and separation from nature. By increasing density and accessibility to green space, issues of mental health, physical health, and sustainability are addressed simultaneously (Friedman, 2022). People need green areas, parks, squares, and gardens, which are numerous and located in such a way that every house and every workplace is within at least 15 min distance (450 m). People can only use them and regenerate their strength when they are close, because people are discouraged by distance. Large scale parks are needed for longer moments needed to regenerate after work (Trojanowska, 2023). They can be equipped with playing fields and recreational infrastructure and used for organized mass events. It is important to think holistically about the city and green public areas in a hierarchical approach.

Grid of green connections

The green and blue infrastructure is formed by public parks and private gardens connected by a grid of green connectors. They are composed of linear parks, green streets, green walls, and green roofs. Green streets have green lines along pedestrian routes or traffic lines. The tree-planted alleys offer additional advantages of tree canopy. Cecil Konijnendijk (2023) proposed a 3-30-300 rule for urban forestry and greener cities. The first element of the rule is that every citizen should be able to see at least three trees from their home windows. The second is to guarantee 30% tree canopy cover in every neighbourhood. The third one is to guarantee every urban dweller the maximum distance of 300 m from the nearest park or green space. The green connections make the difference when it comes to walkability. Green, tree-lined streets with tree canopy and places to sit and rest are inviting people to walk to local public parks and open green spaces. There must exist a perception of achievability in the destinations they are aiming to reach, for people to use green spaces on a daily basis. Moreover, the walk should be a pleasant experience (Alfonzo, 2005). To encourage inhabitants to reach outdoor spaces using active mobility, the grid of pedestrian-friendly green streets would be beneficial. The grid of green connections is also important for stormwater retention and infiltration. Permeable pavement can be effective, as it allows water to seep into a reservoir below.

The grid of green connections in dense urban conditions is filled with green walls and green roofs. Green walls usually take the form of one of two arrangements: green façades or living wall systems. Green façades are composed of climbing plants, e.g., vine plants climbing from the natural soil or potted climbing plants located on balconies or terraces. The living wall systems are modular systems, which allow immediate covering of large surfaces. Some of them are equipped with dripping hydroponic irrigation systems that cycle water from the top to the base and back. Green roofs allow for rainwater detention, improve microclimate, perfectly filtering air pollutants, and are an invaluable source of oxygen. They provide additional thermal insulation and protection against noise. Green roofs can be covered with extensive vegetation, i.e., purely decorative, or intensive – walkable, thus more demanding, but always leading to increasing biodiversity. Buildings with green roofs should be low rise, within the operation range of majority of flying organisms. The closer to the ground, the better connection to other ecosystems. Green walls can be beneficial when the roof is also green, establishing a path between the natural soil level and miniature habitats that blanket built forms (Friedman, 2022). Ecological continuity also means the possibility of providing replacement habitats for individual species and strengthens the resilience of the green and blue infrastructure. Green roofs are often a way to regain biologically active space on the investment site.

Biodiversity protection

Biodiversity provides the main basis for ecosystem services that are important to human life and well-being, like reduction of urban stress and long-term health effects (Feliot-Rippeault, 2023). The dramatic declines in biodiversity which are reported recently can hinder the psychological benefits of nature on our mental health.

Biodiversity needs connectivity and ability of organisms to relocate across multiple habitats, expanding their communities and allowing interactions between those in separate spaces. Smaller ecosystems can support each other and strengthen biodiversity when they are connected into a network of green open spaces (Friedman, 2022). Planning for biodiversity should begin at a larger scale and link green areas. If dangers of decreased biodiversity are left unrestrained, they have the potential to become much worse (Friedman, 2022). Biodiversity protection requires mitigation of the effect of human disturbances and pollution.

Biodiversity protection means promotion of local and native species. Stable and properly functioning ecosystems need pollinating insects, thus introduction of melliferous plants for pollinators is important. It is important that such plants bloom from early spring to late autumn, ensuring pollinators in urban areas have constant access to food.

The cultivation of local biodiversity provides physical and psychological benefits for people. The psychological benefits increase with the species richness of urban greenspaces (Fuller et al., 2007).

Re-naturalization

Renaturation could be defined as recovering a space of nature such as it was in its original indigenous state (Richard, 2023). It can have a doubly beneficial action for human life and for biodiversity by allowing nature to regain its place. Renaturation could be understood as the restoration of meaning and social ties in degraded anthropized environments through the reintegration of floristic, faunistic, and human life (Richard, 2023). Re-naturalization may also be understood as restoration of natural habitats in biologically degraded, polluted, or contaminated areas. The possibilities of modern technology are increasing, but the most effective is phytoremediation: the use of plants for the soil or water purification process. Re-naturalization of the habitat may be translated into the protection of biodiversity. The principles of phytosociology and the laws of species succession are important to shape urban conditions with nature not against it. Long-term thinking is recommended when selecting species for planting in public spaces. Striving to restore natural habitats is the best way to reduce artificial maintenance methods and expenses.

Roderick Lawrence (2020) stipulates that there are two approaches to re-naturalization. The first involves provision of greenery, irrespective of energy, water, and financial costs, e.g., biological artefacts, vertical gardens, and high maintenance public parks. The nature serves as a superficial embellishment, not an essential force which permeates the city. Hence, the promises to improve living conditions are generally false on the long run and in larger scale. The second is systemic approach based on working with nature. It applies relational thinking along principles of circular economy and interprets urban regions as metabolisms (Lawrence, 2020).

The pursuit of sustainable development requires the abandonment of the use of chemicals – fertilizers, pesticides, and herbicides. It also means using local plant species and limiting exotic varieties. It is difficult to give them up completely when they are needed to create beautiful garden compositions, but we can try to limit the number of varieties that require expensive care. It is necessary to give up invasive species that are dangerous to local flora and fauna, but also plants that are highly allergenic and poisonous. More and more often, ecologically expensive lawns are replaced by flower meadows that do not require fertilization or frequent mowing. Moreover, they are an attractive source of food for insects, especially bees. In eco-neighbourhoods, the natural mowing is done by gentle, human-friendly grass-gnawing animals – sheep and goats. This reduces the consumption of crude oil and its derivatives in petrol lawnmowers. Manual mowing is also used in selected locations.

Native plant communities

The tradition of planting decorative flora that require large amounts of resources for their maintenance should be stopped. The use of artificial chemicals, fertilizers, herbicides, and pesticides is not only unsustainable but harmful to surrounding habitats and ecosystems. Native plants are naturally suited to the local conditions and require only minimal additional maintenance, water, or natural fertilizers. Young seedlings may require more water initially before they develop stable root systems and establish themselves. Mature native plants are more drought resistant. Sometimes, however, designing the urban greenery with use of native plants only may require extensive activities to educate potential users. In colder climates with shorter vegetation periods, the native plants may be interesting, but usually do not have large, colourful flowers. Accepting re-naturalization and restoration of natural ecosystems typical of a given habitat may be a challenge. People expectations are based on experiences and tradition of decorative gardening. The sustainable solutions are accepted, if the population is well-informed why they are used and what are the ecological benefits. Replacing exotic ornamental plants with native species decreases or eliminates the need for maintenance, power consumption, chemical applications, and water usage, creating more sustainable landscapes. The local flora and fauna can re-create native ecosystems.

A disturbing phenomenon is the globalization of the ornamental plant market that can be observed all over the world. A few of the most popular ornamental species are used in garden design all over the world (Ignatieva, 2012). The same, fashionable species of ornamental plants are planted in various climatic zones on different continents, regardless of the requirements and natural conditions of a given habitat. Some of these plants may be invasive species, others require high maintenance with artificial irrigation, fertilizers, and pesticides to grow in a habitat far from their indigenous original conditions.

Invasive species

Introduction of non-native species is a risky business. Many plant species used as ornamental plants are brought by humans from other parts of the world. The reasons were their decorative values and other features that were particularly desirable in urban greenery plants, resistance to environmental conditions, rapid development, large weight gains, easy regeneration and spread. However, the same features cause their expansion into areas of natural, native ecosystems, where local species have no chance against them in the fight for survival. Many cases of uncontrolled spread of invasive species can only be observed many years after the introduction of the first individual to the ecosystem. Plants, which do not pose danger to the environment in native habitats, may become invasive in other ecosystems, and multiply without limits eliminating local flora. This sometimes leads to the disintegration of entire ecosystems and a threat to biodiversity. Invasive species create extensive patches of monoculture. While the native ecosystem ensures constant food production due to its wide range of species with different flowering times, the invasive species monocultures' nutritional usefulness for native species, e.g., pollinators, is limited. Additionally, alien plants pose a potential threat in terms of specific microorganisms that are dangerous to native, immune-free species. We also have no data on their longevity in a foreign habitat. We cannot predict the far-reaching effects of plant succession of invasive plants.

Many of the typically ornamental species are not used by the local fauna, which suffers from a deficit of habitats suitable for nesting and food base. The recommended solution is to replace typically ornamental species with native, habitat-appropriate species of vegetation, trees, and shrubs.

Restraining chaotic urbanization, smart growth

Reclaiming previously developed land and adapting it to new uses reward sustainable development. It limits and sometimes even eliminate the cost of construction of new infrastructure, including roads, electricity, water, heat energy, gas, and other services. Building on previously developed areas may involve higher costs, property acquisition, urban taxes, site remediation, removal of existing structures, disturbances or relocation of existing infrastructure, etc. Developing agricultural land is easier and cheaper for the developers, but it is extremely costly for the new dwellers, municipality, and the environment.

There are various options: development of empty spaces in built-up areas, reclaiming previously urbanized and polluted areas, or adapting them to new functions and reuse of historic buildings.

There should be a certain order of urbanization. The agricultural land with productive soil should be saved from development, even if it is located within city limits. It may be used for allotment gardens, urban agriculture, or green infrastructure.

Only the land which is unsuitable for agricultural purposes should be developed in an organized manner according to the rules of sustainability. One of the priorities should be the continuity of green and blue infrastructure, undisturbed air movement through urban ventilation corridors, and preservation of therapeutic landscapes. The beautiful natural landscapes and scenic views should be preserved as a common good. The development of public parks should be given a priority in health-promoting scenery. Chaotic development which is spoiling natural beauty of landscape should be restricted.

Any development must be planned around the natural topography, as modifying it may cause irreparable damage to the existing ecosystem (Friedman, 2022). Modifying natural drainage may hinder the water flow and cause inundations. Any alterations should be minimized.

Smart growth is adapting multifaceted approach to planning, to offer interdisciplinary perspective on city and neighbourhood development. There are several advantages of smart growth including effective use of existing infrastructure, reducing urban sprawl, and decreasing and repairing environmental damages. Smart growth facilitates diversified housing options and align with social well-being. The smart growth strategy departs from master planning (top-down approach) to more organic bottom-up allowance for private commissioning and public participation.

Urban sprawl

The location of a new development is a fundamental decision that affects the sustainability of a neighbourhood. The consequences range from destruction of natural environment, long commuting distance, fuel consumption, and pollution to health of inhabitants and users. Urbanizing undeveloped land in the outskirts of the city brings many negative effects to human health. Those who are living in low-rise suburbs far away from the city centre, workplaces, educational and commercial services may suffer from increased obesity rates, sedentary lifestyles, chronic noncommunicable diseases, and reduced access to healthcare facilities. Very low density will not justify financial investment in public transit, local commerce, and other amenities. Urban sprawl promotes reliance on private vehicles. In places with low density housing, there is an increased rate of obesity due to a lack of physical activity.

Reducing urban sprawl requires addressing challenges such as unplanned urbanization, inadequate basic infrastructure, lack of basic services, and extreme air pollution. When deciding on a location for a new project, there are three possibilities: greenfield, infill, and brownfield sites. Greenfield development, even though it is the cheapest and easiest for developer, has the highest toll on the environment. It contributes to urban sprawl. Infill sites, inside urbanized areas, rely on proximity to utilities and services; however, the zoning ordinance and lot size may be problematic. The brownfield requires extra caution as the site may be polluted and requires remediation. The most sustainable is to increase the density and minimize urban sprawl.

A new investment located outside an urbanized area requires the construction of access roads and utilities, e.g., water, electricity, natural gas, and sanitary sewage networks. This entails not only economic but also environmental costs. Residents must commute to jobs and services, which means moving around and using non-renewable fuels. Therefore, investments within the developed urban tissue, where there are public roads, developed infrastructure, and public transport network, are less burdensome for the environment. It is recommended to strive for the development of wastelands, revitalization of degraded areas, and reconstruction of existing facilities to prevent the urban tissue from spreading into suburban areas, especially agricultural ones (Trojanowska, 2020).

Urban and rural perimeter, green belts

Green belts are continuous stretches of open spaces that are forming part of green and blue infrastructure. They can be used to preserve native flora and fauna. Green belts may act as connectors of separate patches of open spaces, ranging in sizes and forming a hierarchy of public parks. They are integrated with the neighbourhood with linear connectors of green streets, walls, and roofs. The green belts are important for aeration of the city and should be maintained clear of development obstructing natural air flow. The problem with green belt is that they are using a significant amount of land and may reduce the area's density. The provision and maintenance of green belts is a political decision, followed by legal and institutional frameworks for developers, public administrators, property owners, professionals, and other stakeholders.

Reduction of urbanized areas

The pursuit of maximizing the biologically active area requires reduction of urbanized areas. Several actions are undertaken, e.g., compact development, reduction of build-up area, minimizing the impermeable surface, minimizing the vehicle circulation and on-ground parking areas. Multi-story car parks and underground garages allow for additional green areas. Reduction of build-up area means limiting the ground floor area, adding additional floors, or elevating the architectural structures.

Compact development means higher building density. The term "compact development" also refers to a type of quarter development in which the frontages of the buildings fill the entire front of the quarters, without leaving any gaps. It leads to maximum use of the land area in relation to the building area. Compact development means shortening the distances between buildings. Compact communities are coherent, sustainable communities. Compactness allows for better accessibility to basic everyday services and recreation areas because they are located closer to residential houses. Everyday needs can be met by walking. It facilitates the cohabitation of people of different age, income level, or education. Additional benefits include: increasing the dynamics of the economic activity of local entrepreneurs, creating local identity of places, strengthening social ties and safety thanks to "passive observation", and reducing the need to use a car.

Minimizing the impermeable surface is important for enabling infiltration and retention of rainwater, limiting heat islands, maximizing the biologically active surface, and protecting biodiversity. The impermeable surfaces are asphalt roads, sidewalks and paths, sports fields and playgrounds, but also roof surfaces. They can be replaced with grass, gravel, sand, and innovative permeable surfaces on sports fields and playgrounds. The roof surfaces can be replaced with green roofs.

Underground parking and multi-storey parking preserves open green spaces. Even though it increases the construction costs significantly, the indoor parking is ideal in colder climates with snowy winters. Front-facing garages are expensive and can create a repetitive and impersonal streetscape (Friedman, 2022). It is not recommended in high-density neighbourhoods, as it hinders the placemaking efforts. Regular curb cuts for driveways and entrances to garages reduce safety and accessibility (Friedman, 2022). Line parallel parking and rear parking are better options, as they allow for maintaining open and attractive neighbourhood appeal. The parking lots of individual houses can be optimized into one common shared underground garage or multi-level parking serving the neighbourhood.

New urban morphology

The strive for health-promoting urban development leads to urban experiments. Christian de Portzamparc proposed "open block" (French: l'îlot ouvert) and "macro-blocks" (French: macro-lots). The idea is based on optimizing the individual green spaces of each building into one larger open public space. The idea was first operationalized in 1975 in the design of Hautes Formes development in Paris. The concept of Macrolot was successfully used in the last 20 years in various operations of urban renewal in France, e.g., ZAC (Joint Development Zone) Paris Rive Gauche, ZAC Seguin-Rives de Seine à Boulogne, ZAC Paris Clichy-Batignolles, ZAC Ginko Bordeaux, ZAC Docks de Saint-Ouen, ZAC Lyon Confluences, ZAC Jackques Coeur-Port Marianne Montpelier, etc. (Trojanowska, 2024). Chrisitan de Portzamparc believes that the "open block" is an urban form of the third generation, after the closed city quarters of the Haussmann era, and the abandonment of the street proposed by Le Corbusier. The open block and macro-block allows for maintaining a compact urban fabric and meeting basic needs within walking distance. It allows for optimal lighting of residential rooms and designing attractive views from the windows of rooms intended for people. The open block concept makes it possible to match the forms of new development to existing forms of development in urban renewal projects (Trojanowska, 2020). The open urban block is characterized by the increased presence of green spaces and diversified architectural features. The macro-block is aimed at protecting continuity of green and blue infrastructure, shaping aeration corridors, enabling sustainable drainage of rainwater, and ensuring adequate sunlight and natural light illumination of neighbouring buildings.

Centrality versus dispersion

The question about preferences for centrality or dispersion can be translated into higher density versus lower density neighbourhoods.

There is a vast research evidence suggesting that higher densities are correlated with reductions in well-being and negative health effects. People living in high-density neighbourhoods tend to suffer from psychological stress and mental health issues, including psychiatric illnesses (Boyko et al., 2021). Lower quality of life and less social interactions were also reported (Boyko et al., 2021. The results may be correlated with the fact that many social housing estates are built as high-density neighbourhoods. The high urban density may also indicate high density of social problems, unemployment or low income, crime, and social exclusion. Extremely high-density developments, such as 10- to 20-storey apartment towers, offer little public space and often no private or semi-private outdoor space (Friedman, 2022). The demolition of Pruitt-Igoe high-raised social housing, and the unfortunate history of great ensembles in France, demonstrates not only failure of urban planning, but also social exclusion and ghettoization of socially disfavoured groups. The challenge is to promote not only high density of people, but also high density of amenities and social chances for success.

Urban centrality versus the suburbs may be a factor which represents social image of the inhabitant. The symbolism of status is expressed in several ways, and home and office location is one of them. Central location and high densities may improve well-being. High density of residents, workplaces, services, transport, recreation, and public parks might facilitate social interactions, strengthen social contacts and social ties, and build social capital. The design of compacted high-density neighbourhoods reduces travel time and increases accessibility. The distance between the centre and the edges is shortened. High density may promote walkability and physical activity. Relocating to areas with higher neighbourhood accessibility to public transit decreases miles travelled in a vehicle (Friedman, 2022).

Low density will not justify financial investment in public transit, local commerce, and other amenities. Smaller pool of potential clients would not support local commerce and services. On the other hand, low-rise apartments and townhouses can still provide residents with open spaces and privacy while adequately increasing density (Friedman, 2023). Lower densities may result in increased accessibility to open green spaces.

What is needed is urban planning for sustainability. Well-planned compact cities with accessible green spaces, walkable neighbourhoods, and efficient public transportation systems promote physical activity, reduce air pollution, and enhance overall public health. According to Ann Moudon, the physical activity and health of residents is influenced not only by a dense street network and small building blocks, but above all by a variety of functions (Moudon et al., 2006). Higher density allows for optimization of energy resources and investment in innovative technologies.

New standards for public street networks

After many years of designing streets for the automobile to move large numbers of vehicles as efficiently as possible, nowadays the shift is to design street networks that put people first. The low-density, auto-centric patterns of the 20th century failed to deliver health-promoting environments. Today, the priority is on pedestrian safety and well-being. There are many factors that relate street design with health promotion. Walkable neighbourhoods foster higher level of physical activity among all age groups, including children (Boyko et al., 2021). Insufficient physical activity is one of the ten leading risk factors for death across all income scales worldwide, and a key risk factor in non-communicable diseases (Global Design Cities Initiative [GDCI]). When people feel safe about walking around their neighbourhood, they may walk for leisure and transport more often. Walkable neighbourhoods are perceived as more interesting and better designed for social interactions (Boyko et al., 2021). Streets are public spaces that people use on a daily basis. Streets can allow people to live their public life in a city (GDCI, 2024). They inform a sense of place in each neighbourhood, strengthening local identity and embedding historical and cultural meaning. Streets should be designed as context sensitive, and participatory planning supports long-term stewardship of these spaces (GDCI, 2024).

Mariela Alfonzo (2005) developed the hierarchy of walking needs model based on the Maslov theory of motivation (1954). The foundation of the pyramid of basic walking needs are limits related to feasibility of walking to reach a certain destination. This factor influences the decision-making process for all walking trips. As Alfonso specifies, the feasibility of walking may affect the choice between walking and other forms of transportation for destination trips. When it comes to walking for pleasure (strolling trips), the feasibility factors may affect the choice between taking a walk or not (Alfonzo, 2005). The upper walking needs are related to urban form and include accessibility, safety, and comfort and the highest need is pleasurability. If all the walking needs are meet, people are more inclined to spend more time walking for pleasure, and strolling. Walking must be an enjoyable experience in order to invite people to be more physically active.

Each street should be considered in relation to its potential role in the larger network. The best is to keep urban blocks small or provide shortcuts to offer multiple route options. The street safety is directly related to speed limits. The continuity is the foundation of walkability. Sidewalks, bike lanes, and all travel lanes must be continuous and connected in order to function effectively. Clear paths should offer possibility to navigate the neighbourhood to do all the necessary everyday errands without disruption. Parklets, transit stops, landscaping, trees, and parking spaces allow a street to be adapted to various uses without compromising the walkability. People with special needs and disabilities, children, and the elderly would appreciate street furniture and equipment that serve basic needs. There should be a bench to sit and rest every 200 m. Public toilets should be present, well-maintained, and free of charge. Sinks with running water and drinking fountains with mineral water are appreciated by all age groups. Street furniture design and locations should meet the desirable street activity patterns and needs.

Street landscaping and trees integrate green and blue infrastructure and provide accessible contact to nature that can reduce blood pressure and improve emotional and psychological health (GDCI, 2024). Urban trees improve the microclimate, purify the air, and provide shade. Native species improve and support local ecosystems.

Streets can become green connectors of local blue and green infrastructure. They can complement traditional piped water drainage systems. Streets' blue and green infrastructure corridors can capture and infiltrate or evaporate water before it enters the piped system and thus help reduce water pollution and limit the risk of flooding.

The design of public streets network has impact on human health. The success of the street should be measured at human eye level and at walking speed. Pedestrians experience the street with all their senses. Visual clues and interest, textures, sounds, and smells shape the experience and the comfort of the space (GDCI, 2024). Well-designed streets can encourage social interactions and provide space to meet people and feel socially connected. For people with social inclusion challenges, streets should be an inclusive space. Streets design can promote natural surveillance and help build stronger, safer communities. One of the important issues is architectural variety. Streets are like outdoor rooms shaped by multiple planes. The architectural details, facades, windows, setbacks, signs, and awnings define each side of the street.

The smart design of street networks requires utility service infrastructure which is easy to maintain. There are three options for installation of the underground utilities: in the roadbed, adjacent to the roadbed, and in the underground utility tunnel (corridor). The first two options offer savings in land acquisition and reduced construction time. Among the disadvantages are greater space requirements and the fact that repairs may cause disruption to transit, cycle lines, traffic, and temporary loss of pedestrian area. It may be hard to coordinate projects between infrastructure providers and eliminate collisions. The risk for disruption and accidental damage to co-located infrastructure is increased. The third option, the underground utility tunnel, has major advantages: easy access for service, no traffic impact during maintenance, easy coordination between different infrastructures, and lower maintenance cost. The disadvantages are significant initial cost and longer construction time.

The economic aspects of good street network design encompass multimodality and reliability of transit services. Multimodal use encourages people to walk, cycle, and spend more time and money at local retail services. The attractive, safe spaces for transit riders, pedestrian, and cyclists may have positive economic effects and increase the property values. Well-designed streets create environments that entice people to stay and spend time, generating higher revenues for businesses and higher value for homeowners (GDCI, 2024).

Local actions for global future

Pollution of air, water, and soil are a common problem worldwide. Things which in some parts of the world are taken for granted are still not accessible in other places. There are still places in the world which are lacking potable water and sanitation.

Light pollution

Light gives a sense of security and extends the period of use of both outdoor and indoor areas. It promotes security and orientation in space after dark. However, the excessive use of artificial light is causing not only energy wastage, but also light pollution. Excessive illumination greatly disrupts wildlife, disturbs physiological activities, affect navigation systems and natural sleep cycles. Bright illumination muddle processes such as pollination (Friedman, 2022). In the case of plants, lighting is a signal activating flowering and the period of winter rest. Light pollution has an unfavourable effect on human health and natural circadian rhythm. It makes difficult the astronomical observation of the sky. Improper use of external light sources makes the stars simply invisible. It is a pure waste of energy and resources. It affects ecology, economy, science, safety, health, culture, and aesthetics.

Any artificial exterior light source contributes to increased light pollution. Light pollution is caused by lights scattered in the atmosphere, outdoor lighting, indoor lighting, and glare caused by uncovered light sources with high luminance. The most polluting are street lighting, advertising lighting, and buildings illumination (architectural structures and sport stadiums). Some of these lights are unnecessary. They serve neither the safety nor the navigation purposes. The most dangerous are uplight luminaries that, by directing the light stream upwards, illuminate the sky, causing an adverse effect on nature. Modern societies are not going to give up night illumination completely, but should be aware that it must be reduced. The actions to take are to limit the number of light sources and carefully design the luminaires to direct the light downwards with appropriate targeting. The lighting from light poles and fixtures should be focused directly onto the streets. It is recommended to turn off the lights during the night, especially in workplaces, public parks, and gathering spaces. In residential buildings, people can use darkened windows, curtains, and blinds that do not allow light to pass through and help reduce night-time illumination. For outdoor and indoor lighting, we need solutions directing the light stream downwards, with the option of dimming or time limiting it. On streets, pedestrian paths and motor roads, motion sensors and downlights can be used, directing the light stream downwards rather than uplighting.

Air pollution – heating and energy production

There are places where pollution of urban air is so heavy that it is necessary to evaluate the trade-offs between ventilating with polluted outdoor air and letting indoor pollutant concentrations escalate. Reduction of pollution sources, both from indoor and outdoor origins, requires political decisions and actions, as well as public opinion support. Elimination of volatile organic compounds (VOC) from the interior building materials is crucial to improve the indoor air quality. Indoor air quality may be conditioned by HVAC systems. Mechanical ventilation systems can be equipped with air purification filters. Airtight building envelope helps to eliminate heat escape through thermal bridges and improves the results of the net-zero energy passive buildings (NZEB). At the same time, natural ventilation has its advantages. It minimizes dependence on energy for ventilation and allows using indigenous vernacular solutions.

Either way, both natural and mechanical ventilation require decent quality of exterior air. Air quality improvement requires actions on a larger scale of city and regional planning. Location of urban ventilation corridors, the continuous grid of blue and green infrastructure, size and shape and open green areas, presence of street greenery, number and quality of mature trees, etc. – all this can significantly improve the microclimate and air purity. One of the major problems of contemporary cities is the obstruction of ventilation corridors. Development of structures blocking open green spaces which serve to deliver fresh air to the metropolitan areas should be stopped.

Soil pollution

Urban renewal projects usually require careful examination of potential site contamination. Industrial sites may be polluted with heavy metals, crude oil, chemical substances, and other pollutants. Site contamination is a challenge, as it can potentially affect the health of future residents. There are various methods, but the one which is the cheapest and ecologically sound is the phytoremediation – elimination of pollutants with use of plants. The drawback of this method is the long time needed to remedy the soil pollution by the plants (Trojanowska, 2022).

Noise pollution

Noise pollution causes a number of health problems: cardiovascular issues, hearing impairment, sleep disturbance, poor work and school performance, etc. One of the primary sources of noise pollution is street noise caused by heavy traffic. Urban planning should transfer the heavy traffic away from residential areas. Reducing speed and horn use may promote comfort for urban dwellers.

Architecture

Friedman (2022) lists six key design principles when planning a sustainable dwelling. He advocates for: minimizing size, simplifying configuration, joining units, stacking floors, dimensioning for modular design, and planning the interior efficiently (Friedman, 2022). Minimizing size lets for maximizing biologically active areas while reducing ecological footprints. Smaller residences are easier to construct and maintain. Smaller homes help to keep larger open green spaces free from development. Simplifying the configuration, joining units, and stacking floors allow for energy efficiency and more efficient heating and cooling. Among other advantages, he mentioned grouping the facilities in similar space, thus limiting the materials required for pipes and ducts. The ratio of floor area to perimeter should be optimized to achieve the NZEB. Modular design facilitates resiliency and adaptability to inevitable changes (Friedman, 2022).

Form

Form follows function should be replaced with form follows the natural conditions available on site. Proper site planning can decrease the future energy costs. The siting of the building, clever use of sun exposure and wind patterns can greatly enhance the energy performance of the building on both hemispheres. For passive energy gains, orienting a structure and its windows to capture the sun will help reduce the energy needed for heating and lighting. On northern hemisphere, the building openings and windows should be oriented towards natural South, only with a limited variances allowed. On southern hemisphere, the building openings and windows should be oriented towards natural North, only with a limited variances allowed.

The knowledge about the wind patterns can help natural cooling and protect from cold winds during cold seasons. All must be taken into consideration for enhancing local microclimate to suit human needs: natural landscape, local plants and trees, solar exposure, topography, riparian areas, existing buildings and roadways. During summer, structures need to block heat while letting in natural light. Deciduous trees planted in front of the sunny facades help to reduce excessive solar penetration with their foliage. They allow around 10%–25% of radiation to pass through when well foliaged (Friedman, 2022). During winter, when there are no leaves they allow the passive sunshine solar energy gains. Large roof overhangs may act in similar way to deciduous trees, allowing the lower angle winter sun's rays to reach the house and blocking the excessive solar heat in summer.

Rectangular-shaped homes are preferred in colder climates. Studies have shown ratios of 1:1.1 up to 1:1.3 are ideal, whereas more humid regions may prefer ratios anywhere between 1:1.7 and 1:3 (Friedman, 2022).

To preserve the landscape scenic qualities, adapting the height of buildings to the urban context and maintaining existing landscape values are important.

It is necessary to protect neighbouring properties from overshading and deterioration of the view from the window. Another important issue are the psychological aspects related to living in tall and high-rise residential buildings. Christopher Alexander (1977) defended the four-story limit referring to evidence showing that tall buildings lead to the destruction of the human psychological well-being. Living in a tall or high-rise building may lead to a feeling of isolation and alienation, and in the long run, destroy social relationships. Looking out of the window on the upper floors, it is difficult to recognize human faces or voices. Leon Krier (2011) drew attention to the problem of specific density associated with high-rise buildings. In urban planning, there is a well-known problem of a cul-de-sac, which becomes a congested junction during rush hours because all residents want to join traffic on the main road at the same time. High-rise buildings are like dead ends, only built vertically. They also contribute to similar problems.

Creating friendly public spaces requires maintaining a human scale – distances small enough to determine the expression of a human face and hear a human voice. It is worth dividing open spaces into smaller, subjective landscape interiors by, for example, planting high and medium-high greenery (Trojanowska, 2020).

Sunlight and daylight

Adequate lighting is important for health and well-being of residents. Work activities and hobbies require sufficient lighting. Poor lighting or uneven lighting can cause eye strain, fatigue, headaches, and irritability. Without sufficient exposure to sunlight, people can experience seasonal affective disorder; bone disease due to vitamin D deficiency; sleep disturbances; and a lack of concentration (Friedman, 2022).

References

Alexander Ch. 1977. *A Pattern Language. Towns – Buildings – Construction*. Oxford University Press, Oxford.

Alfonzo M. 2005. To walk or not to walk? The hierarchy of walking needs. *Environment and Behavior*, 37(6), 808–836. https://doi.org/10.1177/0013916504274016.

Boyko T., Cooper R., Dunn N. 2021. *Designing Future Cities for Wellbeing*. Routledge, London.

Feliot-Rippeault M. 2023. Biodiversity. In: Marchand D. (Ed.), *100 Key Concepts in Environmental Psychology*. Routledge, Abington, Oxon, New York, pp.16.

Friedman, A. 2022. *Designing Innovative Sustainable Neighborhoods* (1st ed.). Routledge, London. https://doi.org/10.4324/9781003203025.

Fuller R.A., Irvine K.N., Devine-Wright P., Warren P.H., Gaston K.J. Psychological benefits of greenspace increase with biodiversity. *Biology Letters*, 3(4), 390–394. https://doi.org/10.1098/rsbl.2007.0149. PMID: 17504734; PMCID: PMC2390667.

Garvin E.,Cannuscio C., Branas C. 2012. Greening vacant lots to reduce violent crime: A randomised controlled trial. *Injury Prevention*. https://doi.org/10.1136/injuryprev-2012-040439.

GDCI (Global Design Cities Initiative). 2024. *Global Street Design Guide*, National Association of City Transportation Officials, New York, available online: https://globaldesigningcities.org/publication/global-street-design-guide/resources/references, retrieved on 19.08.2024.

Hartig T., Mitchell R., de Vries S., Frumkin H. 2014. Nature and health. *Annual Review of Public Health*, 35, 207–228. https://doi.org/10.1146/annurev-publhealth-032013-182443.

Ignatieva M. 2012. Plant material for urban landscapes in the era of globalization: Roots, challenges and innovative solutions. In: Richter M., Weiland U. (Eds.), *Applied Urban Ecology: A Global Framework*. Wiley-Blackwell, Chichester, UK/Hoboken, NJ, pp. 139–151. https://doi.org/10.1002/9781444345025.ch11.

Konijnendijk C.C. 2023. Evidence-based guidelines for greener, healthier, more resilient neighbourhoods: Introducing the 3–30–300 rule. *Journal of Forestry Research*, 34, 821–830. https://doi.org/10.1007/s11676-022-01523-z.

Krier L. 2011. *Architektura wspólnoty*. słowo/obraz/terytoria, Gdańsk, Wydawnictwo.

Kuo F. 2010. *Parks and Other Green Environments: Essential Components of a Healthy Human Habitat*. National Recreation and Park Association, Ashburn, VA.

Kuo F., Sullivan W. 2001. Environment and crime in the inner city. Does vegetation reduce crime? *Environment and Behavior*, 33(3), 343–367.

Kuo F., Bacaicoa M., Sullivan W. 1998. Transforming inner-city landscapes. Trees, sense of safety, and preference. *Environment and Behaviour*, 30(1), 28–59.

Lawrence R.J. 2020. *Creating Built Environments: Bridging Knowledge and Practice Divides*. Routledge, Abingdon, Oxon, New York.

Louv R. 2005. *Last Child in the Woods: Saving Our Children from Nature-Deficit Disorder*. Algonquin Books, New York.

Maas J. 2008. *Vitamin G: Green Environments – Healthy Environments*. Nivel, Ultrecht.

Maas J., Verheij R.A., de Vries S., Spreeuwenberg P., Schellevis F., Groenewegen P.P. 2006. Green space, urbanity, and health: how strong is the relation? *Journal of Epidemiology and Community Health*, 60(7), 587–592. https://doi.org/10.1136/jech.2005.043125.749.

Mass J., van Dillen S.M.E., Verheij R.A., Groenewegen P.P. 2009a. Social contacts as possible mechanizm behind the relation between green space and health. *Health and Place*, 15(2), 586–595.

Maas J., Verheij R.A., de Vries S., Spreeuwenberg P., Schellevis F., Groenewegen P.P. 2009b. Morbidity is related to a green living environment. *Journal of Epidemiology and Community Health*, 63, 967–973.

Mitchell R., Popham F. 2008. Effect of exposure to natural environment on health inequalities: An observational population study. *Lancet*, 372(9650), 1655–1660.

Montgomery Ch. 2015. *Happy City: Transforming Our Lives Through Urban Design*. Penguin Books Ltd, London.

Moudon A.V, Lee Ch., Cheadle A.D., Garvin C., Johnson D., Schmid T.L., Weathers R.D., Lin L. 2006. Operational definitions of walkable neighborhood: Theoretical and empirical insights. *Journal of Physical Activity and Health*, 3(Suppl 1), S99–S117.

Richard I. 2023. Renaturation. In: Marchand D. (Ed.), *100 Key Concepts in Environmental Psychology*. Routledge, Abington, Oxon, New York p. 110.

Trojanowska M. 2017. *Parki i ogrody terapeutyczne*. Wydawnictwo Naukowe PWN, Warszawa.

Trojanowska M. 2020. *Poszukiwanie standard projektowania ekoosieldi w Polsce*. Wydawnictwa Naukowe Uniwersytetu Technologiczno-Przyrodniczego w Bydgoszczy, Bydgoszcz.

Trojanowska, M. 2022. Parki i ogrody terapeutyczne w służbie zdrowia psychicznego. In: Gawrych M. (Ed.), *Natura a zdrowie psychiczne, Wydawnictwo Akademii Pedagogiki Specjalnej im*. Marii Grzegorzewskiej, Warszawa pp. 151–177.

Trojanowska M., 2023a. *Projektowanie zielonych przestrzeni publicznych*. Wydawnictwo Naukowe PWN, Warszawa.

Trojanowska M. 2023b. Reclamation of polluted land in urban renewal projects. Literature review of suitable plants for phytoremediation. *Environmental Challenges*, 13, 1–7. https://doi.org/10.1016/j.envc.2023.100749.

Trojanowska, M. 2024. The evolving theme of health-promoting urban form: Applying the macrolot concept for easy access to open public green spaces. *Urban Science*, 8(3), 115. https://doi.org/10.3390/urbansci8030115.

Zhang L., Tan P., Diehl J. 2017. A conceptual framework for studying urban green spaces effects on health. *Journal of Urban Ecology*, 3(1), 1–13. https://doi.org/10.1093/jue/jux015.

Zhang X., et al. 2020. Indoor air design parameters of air conditioners for mold-prevention and antibacterial in island residential buildings. *International Journal of Environmental Research and Public Health*, 17(19), 7316. https://doi.org/10.3390/ijerph17197316.

4 Economy

Potable water

The problem of drinking water is acute all over the world. Reducing drinking water consumption and sewage discharge are among the pillars of sustainable development. There are various recommendations on how to reduce the consumption. Firstly, the education of public about pro-ecological attitudes and changing the habits, e.g., turning off the water tap when brushing teeth or shaving, taking a quick shower instead of bathing in a full bathtub, and rinsing vegetables and fruits in a pot instead of under an open stream of water. Installing individual consumption water metres may help to calculate the real usage per capita. The attitude should be to take as little water from nature as possible.

Secondly, limiting the use of drinking water for non-consumptive purposes, e.g., touchless faucets, thermostatic faucets, faucets with aerators, and tank flushers mounted to toilet bowls with smaller water capacities and equipped with double buttons.

Thirdly, storm water retention and management could help to limit the potable water use. Permeable urban surfaces allow for rainwater retention and infiltration.

Fourthly, promoting xeriscape landscaping with native plants to reduce the need for watering.

There are several problems with drinking water resulting from pollution: microplastics and heavy metals. Water shortage is causing increased energy consumption. Pumping water each time from deeper wells needs more and more energy.

Reduction of water use

Potable water consumption may be limited if it is reused. The wastewater or grey water reuse for flushing toilets requires dual installations – one for potable water only and the second one for process water (purified grey sewage): sewage with low content of pathogens, e.g., rainwater, sewage from sinks, bathtubs, and washing machines, which can be purified and reused within 24 hours. Grey sewage can be divided into slightly polluted (light grey: bath, washbasin) and heavily polluted (dark grey: sink, dishwasher, washing machine).

DOI: 10.4324/9781003494898-5

There are also known examples of the use of heat recovery installations from grey sewage to support building heating.

The attitude should be to create recycle loops as small as possible: regenerate, recycle, e.g., energy from water and nutrients from water. It is technologically possible to purify wastewater to the standard of drinking water. The challenge may be socio-cultural barriers to using recycled water and acceptance of water reuse.

Xeriscaping

Landscape development with native sturdy plans minimizes the need for watering and maintenance. Proper selection of vegetation in combination with certain design strategies can lead to beautiful site design without excessive irrigation, polluting chemicals, and energy use.

Another option is soil amendments, adding additives to soil. Example of such additives is mulch, beneficial in retaining water, maintaining low soil temperatures, and hindering erosion processes (Friedman, 2022).

Storm water retention

Much of the 20th century, the paradigm for storm water management was to discharge it from the site as quickly as possible and transport it with underground pipes to local receiver – pond or lake. The rapid outflow of rainwater directed by storm sewage systems to distant recipients may cause local flooding during heavy rain, and other extreme weather phenomena, and thus lead to the depletion of groundwater resources. Nowadays, we are facing changes in paradigm towards nature-based solutions. Striving for sustainable development means ensuring the natural circulation of water in nature. In addition to minimizing the built-up and paved surface, solutions are used that inhibit too rapid surface runoff.

A drop of water should soak into the ground where it falls, to supply the underground water basins and prevent drought. If it is not possible in densely urbanized areas, it is recommended to collect rainwater into tanks and use it for washing, flushing toilets, washing roads and sidewalks, watering house plants, and other household needs.

Individual water retention devices (raingardens or tanks) are envisaged for private gardens. In the neighbourhood scale, rainwater can be managed using surface retention systems that ensure slow infiltration of water into the subsoil layers in the place where the drop fell on the earth's surface (sustainable urban drainage systems [SUDS]). SUDS system should be combined with blue and green infrastructure grid and street layout to provide a comprehensive strategy of rainwater management. The system of devices used on the ground may be complemented with green retention roofs and walls. SUDS is land consuming but contributes to improved microclimate, reduction of heat islands, and increased biodiversity. The presence of plants in a rain garden plays a very important role because it helps to remove impurities and retains water. It is important, especially during draught periods, as it increases air humidity and improves the microclimate.

Permeable urban surfaces

Parts of the street system – roads, walkways, driveways to buildings or garages, parking spaces, sidewalks, and paths – require hardening to perform their function properly. Impermeable surfaces, especially applied to large surfaces and dark coloured, are responsible for heat islands within cities – local areas of increased heat during the hot days, and depletion of potable water resources. Application of permeable pavement around buildings allows rainwater to seep into the ground to filter to the underground water basins. There are some innovative solutions which ensure hardness, strength, and durability, as well as permeability of surface. The layers of gravel and sand are laid without the traditional impermeable cement ballast.

Energy

The global consumption of energy is constantly increasing. What is needed is a major shift towards renewable energy and decreasing the overall energy use. What needs to be restricted is the primary energy dependence. Primary energy, often called environmental energy, is the energy contained in non-renewable fuels, necessary to cover the demand for final energy supplied to the building. Primary energy consumption is calculated taking into account the efficiency of the entire chain of production, acquisition, conversion, and transport. Final energy is the amount of energy supplied to the building in order to ensure the provision of calculated air temperature in the rooms, an appropriate ventilation, and hot water. The efficiency of the heating, cooling, and hot water preparation systems is taken into account. Final energy approximates the energy costs incurred by the user and indicated in the energy performance certificates of buildings and apartments.

In the old model, energy flowed in one direction, from the power plant to the users. The development of renewable energy sources results in the need to implement bidirectional energy flows. It consists of many different energy production sources along with energy storage facilities. Generation sources should be able to operate in a separate network and be mutually redundant. The new model of energy production and distribution is based on decentralization of services to eliminate loses during the transportation and ensure security of supply. Distributed energy production allows for inclusion of new low- or zero-emission generation sources limiting the use of fossil fuel resources. In this concept, energy transfer can be bidirectional: both in the traditional direction (from large energy generation to individual consumers, from the highest voltages to the lowest) and in the opposite direction (from prosumers to distributors). The new model of energy production stipulates creation of local smart grids. The related energy flow management systems will be located in municipalities and industrial plants at individual recipients, especially prosumers. It is necessary to implement bidirectional intelligent measurement systems, e.g., bidirectional energy meters with the function of forecasting energy consumption over time.

The diversification of energy sources and energy self-sufficiency are a great value both on the macro scale – country, region, and city, and on the micro scale – eco-neighbourhood and individual building. The aim of diversification is to reduce the risk of disruptions in energy supply. In the case of renewable fuels, it may be difficult to store energy in periods when its production is higher (summer), and photoelectric conversion when the intensity of its production decreases and demand increases (winter). Sufficiency is the ability to cover customer needs at all times. The shift towards renewable energy is gradually implemented in large and small scales. There are several ways to produce energy without relying on fossil fuels, such as geothermal, biomass, refuse-derived fuel (RDF) (combustible non-recyclable waste), and solar and wind energy. On the neighbourhood scale, the district central heating, cooling, and energy production are offering energy efficiency and cost savings resulting from the economy of scale. It is easier to implement ecological and innovative solutions for a group of users with a collective budget and larger land area, than one investor on an individual plot. Investments in microgrid, micro cogeneration, micro wind turbines, and energy storage are more feasible for a group of users. Areas comprising multi-functional and dense structures are of special interest for district energy production, heating, and cooling. Mix of functions affects the temporal patterns and reduces the peak load. The efficiency of district systems can vary depending on urban density and demand. When consumers are located in close proximity, it increases the efficiency of the network. The district energy, heating and cooling systems should be planned for future expansion.

Energy efficiency

Energy-saving improvements, additional thermal insulation, reduced dampness, and affordable warmth have positive impact on health.

The most important for energy efficiency is the energy awareness of users. Heightened energy awareness can lead to changes in the way energy is used, reducing overall consumption (Friedman, 2022). Switching the appliances and turning the lights off when leaving the room may become a habit and lead to long-term energy reductions.

New facilities should be built in a passive building standard, using renewable energy technologies for heating and, if necessary, also cooling. Existing buildings can be thermally modernized and remodelled to bring them to the passive building standard. Minimizing energy needs for heating and cooling combined with the production of renewable energy allow for energy self-sufficiency.

The objective for energy-efficient urban planning is investing in areas where limited energy consumption is possible. Urban revitalization of brownfields allows for significant savings in embedded energy, which is consistent with the demands of sustainable development. Therefore, today it is no longer advisable to demolish buildings, which are still technically efficient, but to choose the options of revitalization and remodelling. A conservative approach prevails, so as not to waste material and energy resources built into existing buildings. During revitalization, not only the functions but also the appearance of the buildings changes. Post-industrial buildings are adapted to perform cultural, residential, and office functions. Office buildings are being remodelled into residential buildings. The aim is to create buildings that can be easily adapted to the changing needs of users. This approach makes it easier for them to be used by subsequent generations.

To limit the final energy cost, the old buildings need to be retrofitted. They need the newest innovative technologies to be transformed into nearly-zero energy building (NZEB). They would require improvements of thermal insulation of walls, roofs, and substructures to eliminate thermal bridges, waterproofing, and windows replacement for triple-glazed ones. Some of the strategies would require remodelling and installation of mechanical HVAC systems for heat recuperation.

In case of infill development, designing a new eco-neighbourhood with NZEB buildings gives opportunity to use the advantages of site development to increase passive energy gains for heating and cooling. The pursuit of the most optimal use of free forces of nature for heating and ventilation of buildings stipulates careful urban, architectural, and landscape design. Passive gains from solar energy require the implementation of solutions at the design stage, e.g., orientation of the building and location of glazed openings from the sunny side. The parts of the buildings located far away from the sunshine should have no openings.

The largest economies are on a neighbourhood scale. If all homes within a neighbourhood are well insulated and sealed to reduce heat loss, the neighbourhood will also use less energy (Friedman, 2022).

The possibilities of producing and using energy from renewable sources are limited; therefore, the primary goal should be to reduce the demand for usable energy. Strategies in passive construction are selected very carefully and correspond to the detailed climatic conditions of the area, in which this facility will be implemented. In practice, this also means that all solutions, starting from the location of the building, architectural form, number and position of glazing, details of construction, building elements and materials used, and individual installations are supported by numerous calculations.

According to the Passive House Institute, passive house requires less than 15 kWh/(m^2yr) for heating or cooling of the living space. Conventional primary energy use may not exceed 120 kWh/(m^2a) (IPHA International Passive House Association). NZEB are buildings which require nearly zero or literally zero energy for everyday use.

NZEB buildings must meet the following passive building standards:

- compact shape, i.e., minimized ratio of external wall surface to volume,
- windows with appropriate parameters should be placed on the sunny side (e.g., triple-glazed)
- it is recommended to avoid designing windows on the shaded side,
- low heat transfer coefficient of external partitions,
- elimination of thermal bridges,
- airtight and sealed building envelope, and
- the use of mechanical ventilation with recuperation, i.e., heat recovery (heat recovery in ventilation at the level of 85%) (Trojanowska, 2020).

The compact shape is beneficial for energy efficiency of air conditioning: heating and cooling. The compactness of the building is described by the building's compactness ratio (Form Factor) A/V, i.e. the sum of all surfaces of its envelope, divided by its gross air conditioned (heated) volume. Reducing the A/V ratio is achieved primarily by designing buildings with a simple shape, without bay windows, recesses and protrusions in the walls and roof. The compactness ratio (Form Factor) does not depend on building's orientation, whether the building is insulated, or on how much. The compactness ratio (Form Factor) only depends on the geometry of the thermal envelope. Poor architectural design might determine that the building would not meet the passive building standards regardless of insulation (Bonilauri, 2015).

The design of NZEB building requires assessing the building's shading in two seasons: summer and winter. Extensive open spaces in front of the solar façade of the building should be left appropriately to ensure full illumination of the building. In the summer, shading may be beneficial because it prevents the building from overheating. Therefore, planting a row of deciduous trees on the sunshine side – true South (Northern Hemisphere) or true North (Southern Hemisphere) will create a barrier against excessive radiation with the foliage in the summer. The trees with no leaves will let the sun's rays through in winter. Winter shading of walls and windows is unfavourable because it prevents the passive use of solar energy.

The optimal positioning of the building plot, when there are no other factors (e.g., topography) may be determined by cardinal directions: North-South Axis with minimal deviation of 15°–30° from the East-West axis, at which solar radiation is still beneficial to the building's energy balance.

Installing modern heating systems is effective only when unwanted air flows through external partitions are eliminated. As a result, air tightness becomes a condition for meeting current energy requirements. Thermography and air tightness tests of the building confirm the quality of construction works. Thanks to this method, it is also possible to diagnose serious damage to the building structure caused by warm and humid indoor air penetrating the building structure through cracks. Passive houses must be airtight with air change rates being limited to $n50 = 0.6$ 1/h (IPHA International Passive House Association). Minimizing heat loss through partitions requires the use of insulating materials with the best parameters (low values of the thermal conductivity coefficient λ W/m^{2}K) and energy-saving windows (triple glazing). Appropriate insulation of the window frame will significantly reduce thermal bridges, created during the assembly process. A thermal bridge is a place in the building's thermal envelope where thermal conductivity is much higher than that of the partition. All thermal bridges need to be sealed.

There are also other strategies that may enhance the energy efficiency. Among them is temperature zoning and buffering. The use of passive solar energy gains is related to buffering rooms requiring constant temperature by rooms not intended for permanent human stay. Rooms that require the highest temperature should be located from the insulated side or inside the building. Rooms where the temperature may be lower can be designed facing the shaded side. Atriums and greenhouse structures of winter gardens may help to improve the microclimate inside the building and harness passive solar energy gains. The passive solar gains require the design of windows and glazing on the facade, which allow solar radiation to penetrate into the interior, where it is absorbed and stored in the walls and floor made of heat-storing materials. The greenhouse (winter garden) can become a direct passive profit system – both a buffer and a collector if it is on the sunny side of the building. Transparent walls allow thermal energy to penetrate, and the ground is the energy storage element. The direct profit system has the highest instantaneous efficiency among all passive systems. The increase in room temperature is consistent with the increase in solar radiation, resulting in very large temperature fluctuations during the day. Reducing temperature fluctuations is possible thanks to storage systems with high capacity and thermal conductivity. Greenhouses can act as an effective collector in the winter, but in the summer they require protection against excessive solar radiation. The most optimal solution for air exchange in greenhouses is the natural ventilation (Kobylarczyk, 2018; Trojanowska, 2020; Zielonko-Joung & Marchwiński, 2012).

The automation, operation of devices for measuring and regulating heat in rooms, as well as the cooperation of these devices, is a very important issue. The user's accessibility to controllers allowing for individual comfort adjustment (depending on the maintained air temperature) may also result in reduced energy consumption.

Diversified energy sources

Innovative solutions in the field of energy production, enabling the use of renewable energy technologies on a building scale, are as follows:

- solar panels,
- photovoltaic cells,
- heat pumps,
- micro-windmills,
- micro-cogeneration,
- micro biogas plant,
- micro biomass plant, and
- micro biofuel plant.

Solar panels

Solar collectors are devices used for direct conversion of radiant sun energy into heat. In residential buildings, they may be the preferred option, because they facilitate direct savings to the end consumers (inhabitants) pockets. It works by absorbing thermal energy of solar radiation (direct and diffused) through the collector's absorption surface and then transferring this energy to an intermediary medium (water, air, or glycol) that flows through pipe set. The heat transported by the medium is most often used for preparation of hot water.

Photovoltaic cells

Photovoltaic installations are devices that generate electricity from solar radiation. Photovoltaic panels produce electricity which can be used in electrical installations in buildings. The photovoltaic installation can operate in two systems:

- "On grid" – the photovoltaic installation is connected to the power grid, which acts as an energy store. Electricity produced by the installation first powers the installations and devices in the building, and the surplus energy is transferred to the power grid. It can be returned from the grid when the installation is not able to provide the required amount of energy due to temporarily unfavourable weather conditions or greater demand for electricity.
- "Off grid" – the photovoltaic installation is an independent electrical installation, not connected to the power system. It requires connecting the installation to a battery that is able to store the surplus of produced energy. The undoubted advantage of this system is its complete independence from the distribution network and not susceptible to power outages.

Photovoltaic installations work perfectly with heat pumps and/or mechanical ventilation systems with heat recovery.

Heat pumps

A heat pump is a device that changes energy from the environment into heat. They convert energy from renewable sources, such as air, ground, or water, into heat energy used for heating or preparing hot water. Additional energy is needed to power the compressor and pumps. The thermodynamic system of the heat pump allows for very effective use of this energy. It is possible to reverse the heat pump circulation direction to use the same device for both heating and cooling. The thermal efficiency of heat pumps depends on the temperature difference between the lower and upper sources. The most stable solution is to use a ground heat pump, which uses a relatively constant ground temperature during the year. These pumps often perform heating functions in the winter and cooling functions in the summer.

Micro windmills

Wind farms operate autonomously, providing energy instead of downloading it from the power grid. This is possible thanks to the use of special automation in the converter that manages the energy flow between the turbine and the receivers. When demand for electricity exceeds the production capacity of the wind farm, switching occurs to automatically supply the energy from the network.

Smart grid of tomorrow

NZEB and energy-plus buildings can be connected into an extensive network of the eco-neighbourhood (smart grid). Independent energy production requires establishing new purchasing and selling relationships with traditional energy suppliers. The major drawback is the insecurity of supplies. Solar energy depends on weather conditions. Solar energy production takes place mainly during the day, which means it varies in intensity depending on the level of cloud cover. When weather conditions favour high electricity production, an excess problem may arise. The mini energy network is unable to absorb the surplus, which leads to its waste. The surpluses must be sold. Another solution is to use the surplus of produced electricity to charge batteries or electric cars. Temporary energy shortages must be supplemented with energy from larger power plants.

At the neighbourhood scale, it can be proposed to diversify the central production of heat and electricity using, for example, cogeneration or trigeneration, waste incineration plants, or other renewable energy sources. Renewable energy sources include wind energy, solar energy, geothermal energy, biomass, biogas, biofuels, and hydropower. Their unquestionable advantages include inexhaustibility of resources, increasingly lower installation costs, and the possibility of production directly at the consumer's premises. It is also worth paying attention to secondary renewable energy, i.e., fuels obtained from primary renewable energy sources.

Urban metabolism

The waste generated per person per day averages 0.74 kg worldwide, but ranges widely from 0.11 to 4.54 kg. Sixteen percent of the world's high-income population generate about 34% of the world's waste (World Bank, 2024). Rich countries produce more waste than poor ones. In poor countries, things are used for a long time, repaired when necessary, or taken apart to reuse their components. A feature of developed societies is excessive consumption: buying disposable items, unnecessary goods, and quickly replacing existing items with others. The global waste is expected to grow. Thus, the most important part for waste management is education. The consumers' awareness is shaping the attitudes and habits. Example of educational efforts are the 3Rs, 5Rs, or 7Rs principles. There are different variations, but the major concept stays the same. It is worthy to think twice when dealing with material goods. The city of Winnipeg, Canada, explains the 7Rs as:

1 Rethink – your relationship with consumer goods and environment
2 Refuse – products that are harmful to the environment
3 Reduce – the amount of goods you buy and their packaging
4 Reuse – whatever you can and upcycle
5 Repair – fix before throwing away
6 Regift – pass your gifts on somebody else if they don't fit you or sell
7 Recycle – let the used goods enjoy a new life (winnipeg.ca. 2024).

The idea of urban metabolism is related to zero-waste concept, where waste management is closing the loop of the goods' life from cradle to cradle and nothing is directed to the landfill.

To facilitate the urban metabolism and reduce nuisances and odours, the underground waste collection systems are being implemented in larger metropolises. Sometimes, it is combined with pneumatic waste collection systems (vacuum system) (AWCS Automatic Waste Collection Systems). The pneumatic systems can be stationary and mobile. In mobile systems, waste is stored in temporary underground containers under the drop-in points and sucked in automatically and collected by the truck fleet. The vehicle fleet is equipped with a special scale that allows accurate measurement of the amount of waste collected from a given estate.

In the stationary system, the contents of the containers are sucked in by vacuum blowers located in the collective receiving station. Stationary systems involve the construction of a central receiving station within the estate and a network of underground pipelines for waste transfer. At the collective receiving station, the suction machines are turned on several times a day, either at specific hours or after receiving a signal that the containers are full. Monitoring the filling of containers, but also the correct functioning of all system elements (informing about clogging of pipes, etc.) is carried out remotely 24 hours a day using an automatic control system. Waste sent to the collective collection station is compacted using a special press and then successively transported to the target processing sites, waste incineration plants, etc. The entire system is airtight to guarantee hygiene and safety of use. Separate pipelines may be used, e.g., one for bio-waste and the other for other waste, or bag colour detection systems – to facilitate waste segregation at the collection station. The disadvantage of the underground solution is high investment costs and difficulty in integration with the existing technical infrastructure.

Circular economy: zero waste

Circular economy is a model of consumption which extends the life cycle of products as long as possible through sharing, leasing, reusing, repairing, refurbishing, and recycling existing materials (European Parliament, 2023). In practice, it implies reducing waste to minimum.

The zero-waste approach requires support for local services, small entrepreneurship, and cooperatives. This approach involves creating new jobs related to recycling, repairs, and alterations. Workshops, where residents learn how to make basic everyday items themselves, use specialized tools, 3D printers, etc. are gaining popularity. Another phenomenon are second-hand stores and exchange points for, e.g., clothes, books, and home appliances. Motivation is also important, and places where the inhabitants can return used products, e.g., plastic bottles, and metal cans, and receive specific benefits, e.g., cinema tickets, gym memberships, discounts in sports equipment rentals, and discount coupons for the purchase of food products. There are also initiatives to reduce the food waste. Organic produce shops are often supplied by local agricultural and horticultural farms. Zero-waste commerces are encouraging shoppers to bring their own bags and jars. Circular economy reduces raw material dependence.

For sustainable construction the priority is the choice of ecological materials. Reusing existing buildings or their components reduces the impact on natural environment. Production waste actually increases construction costs in two ways: firstly, unnecessary materials (such as packaging) increase the cost of purchased products, and secondly, the cost related to waste disposal.

Local food production

The world today is full of paradoxes. War, hunger, poverty, and environmental degradation persist in spite of knowledge, analytical ability, and newest technology directed towards eradicating them (Lawrence, 2020). Today the obesity coexists along malnutrition (Lawrence, 2020). The industrialization of global food production and major shift from local farming to mass-scale food production alienated the communities from local food produce. They are dependent on national and international agro-industrial companies and have reduced capacity to act effectively in situations of vulnerability. Mass production of food changed human diets and resulted in increased consumption of processed sugar, salt, and saturated fats. These changes are responsible for higher prevalence of obesity and chronic diseases, including cancers, cardiovascular diseases, and diabetes.

During the last century, local food production was slowly eradicated from urban landscape. Gardening, which has accompanied societies in all regions of the world for centuries, was forgotten and rarely encouraged by local authorities and land-use planning projects during the 20th century. Initiatives like urban farming and private or communal gardening are re-localizing food production. There is a movement to promote local food production. The communal food gardens, local distribution of excess produce to populations in need, and reintroduction of farmers markets are tangible outcomes of awareness shift towards food security. Lawrence (2020) defines four characteristics of food security: availability, affordability, utilization, and stability. He declares that people should not risk losing access to food as a consequence of economic crisis, unrest, or other conflicts.

Along with food security comes the concept of food sovereignty. Declaration of Nyéléni (2007) defines food sovereignty as the right of people to healthy and culturally appropriate food produced through ecologically sound and sustainable methods, and their right to define their own food and agriculture systems. The declaration grants the right to conserve and rehabilitate rural environments, fish stocks, landscapes, and food traditions and to defend territories from the actions of transnational corporations.

The concept of food sovereignty is related to permaculture – a landscape that mimics the natural processes to yield abundance of food – and urban agriculture – the practice of food production in urban environment.

Historically, people grew and harvested their own food near their homes. Then it became fashionable to turn land into decorative garden, replacing edible crops with exotic plants with only aesthetic values. Some of those plants turned to be invasive species, escaped from the gardens, spread rapidly, and outcompeted native species. They became a threat to the biodiversity. Today, the practice of edible gardens is slowly returning. They are helping to fight the food insecurities, food deserts, and reduce the risk of food waste. Sustainable development is based on the promotion of organic farming – food produced without the use of artificial fertilizers, genetic modifications, and pesticides. The idea of food security is linked to reduced food miles, i.e., the distance a product must travel from the place of production to the plate. A limitation to overcome is the variable seasonality of fresh products and the acceptance of imperfections in appearance. Gardening in private or community garden can improve physical, social, and emotional health.

Local food production may bring benefits like trade systems with local farmers, which allow manufacturers to sell their food products directly to the final consumer and to other retail establishments. Nowadays the food landscape in the cities is uneven. There are islands with food abundance and food deserts with no fresh food and fast food only. Food deserts are responsible for epidemics in obesity and non-communicable chronic diseases (Boyko et al., 2021).

Local construction materials

The local sourcing of building materials is important for sustainability of construction. It supports the local economy and brings savings. Using materials that can be purchased locally can help reduce energy consumption for transportation. Most building materials can be ordered from local manufacturers or distributors, supporting the local labour market. Specialized construction equipment can be rented from the nearest operator, e.g., construction cranes, concrete pumps, scaffolding, or formwork systems. If possible, it is advisable to create local jobs and employ workers from the neighbourhood who can walk to work or take local public transport. If there is no public transport connection, it is worth considering organizing collective transport for construction workers from nearby towns.

One of the recent trends is to use materials of natural origin. Wood construction is gaining recognition not only in residential but also in recreational construction. Other examples are rammed earth, strawbale, rock and mineral wool, cellulose, hemp, etc.

Local workforce and workplaces

Satisfactory employment and steady income have an impact on health and well-being. Decent income enables access to water, food, shelter, education, and personal development. Unbalanced income causes stress and may lead to mental and physical disorders. Satisfaction with income along with health are important indicators of people's happiness and well-being. Increased income is related to better health. Interestingly, better health is related to a decrease in energy consumption (Boyko et al., 2021). A person's socio-economic status can influence the stress level related to mortality rates (Friedman, 2022). The inability to find employment is one of the most stressful factors in the life of a modern person.

Cities always attracted those seeking employment opportunities. Municipalities which are flourishing economically offer better possibilities for exchanging goods, sharing knowledge and experience, and build a satisfying carrier. Local employment possibilities depend on the economic conditions and functional mix in the neighbourhood. It is a common practice in eco-neighbourhoods to plan and organize workplaces for the inhabitants. The proposed solutions include introducing maximum functional diversity within the neighbourhood, building office space, and providing access to convenient public transport. Extending the housing programme with services creates a small number of jobs in local companies. Additionally, there are various strategies for attracting potential employers with: office space for rent, zones with privileged lease for start-ups, co-working space, and apartments with space for home offices. New opportunities open up by replacing work in the office with remote work that can be done from home. A home office is an important investment for people who are self-employed. Working from home is convenient. The live-work arrangement can help balance career and family life. For centuries people lived in live-work dwelling (Friedman, 2022). There are professions which traditionally work from parts of their own homes (accountants, architects, engineers, lawyers, physicians, etc.), making it possible for professionals to afford entrepreneurship and seek economic freedom, while providing the neighbourhood with vital services. One of the groups which are especially prone to working from home are woman, who are primary caregivers (Kapoor, 2023). The ultimate challenge of working from home is the blurring of distinctions between intimate, personal, social, and public spaces which may affect psychological well-being (Boyko et al., 2021).

The concept of biophilia coined by the biologist Edward Wilson (1984) explained people's preferences for contact with nature rather than the being in a built-up setting. Human being is an integral part of the natural world. That concept stimulated movement to bring nature to workplaces. The view through the window, fresh air, natural light, plenty of potted plants, and bright, happy colours can promote the psychological well-being of workers.

Local employment may also be related to proximity to reliable public transport services. When access to jobs located outside the neighbourhood is facilitated, not only it makes life easier for residents, but also it reduces the additional burden on street traffic caused by individual car transport.

Work gives a sense of independence and self-determination and everyone who wants to take up a professional job should have this opportunity. Employment possibilities should include people with special needs and enable everyone to live a satisfactory life in society. Sometimes the coworking space may also be dedicated to disfavoured groups, e.g., women, young people up to 30 years of age, and people at risk of unemployment after 40 years of age. The design of office space and coworking station may rise some challenges. People need quiet, private space to focus on their jobs. The private workspace to work quietly can be alternated with hot-desking and non-territorial workspace, but the lack of privacy and noise may hinder the productivity (Boyko et al., 2021).

Local commerce and services

The resilience of a community is related to diversified sources of revenue. Municipality should support the local economy to provide products and services generating income and employment. Local businesses create local jobs. Shopping locally encourages economic growth and facilitates sustainability (Friedman, 2022). Sustainable economic development depends on creating a productive, satisfying workplace close to where people live. The entrepreneurship and creativity of residents are crucial for the proper functioning of eco-neighbourhoods. People appreciate the opportunity to work close to where they live, without long commuting hours. There are various advantages of vital business life in the neighbourhood. It boosts the economy, reduces travel to central business areas, and makes the district safer. The streets are watched 24/7 by stakeholders who run their businesses and live in the same community (Kapoor, 2023).

Proximity of basic services is a factor influencing the quality of life. Modern eco-neighbourhoods are designed to promote functional diversity and self-sufficiency. For the basic everyday needs, a full functional programme is provided within walking distance from residences and workplaces. It is assumed that in optimal conditions, residents will work, study, obtain necessary supplies, and relax in the eco-neighbourhood. Other specialized needs can be met using reliable public transport. The public transport services and the location of public transport stops are intended to facilitate satisfying other needs, e.g., advanced healthcare, university education, and philharmonic concerts. (Trojanowska, 2020).

Transportation

Transport enables economic and social connections. Easy access to urban transportation infrastructure is a must for sustainable neighbourhood. Public transport stops should enable affordable, reliable, and quick access to places where various types of higher-level services are located, e.g., hospital, theatre, opera, or university. Transport planning must take into account larger perspective of city life. The key is the integration, not only spatial and temporal, but also functional.

The design of modern eco-neighbourhoods is aimed at reducing car traffic. Instead, the ecological options are proposed: attractive walking paths, bicycle paths, and accessible public transport. There are often some incentives involved, e.g., electric car charging, encouraging people to carpool instead of driving alone, or car sharing.

Active means of transportation are promoted as sources of well-being and health improvements.

The problem of modern cities is the traffic and congestion. Jon Lang (2017) listed three solutions to the overcrowded city centres. First, the standard one is to create more of even wider streets and parking facilities. The second is to improve mass transit system and guarantee high-quality public transportation. The third alternative is to charge people for driving into the city centre (Lang, 2017).

There is a paradox of correlation between car ownership and deprivation (Boyko et al., 2021). In areas of low deprivation, the car was the dominant mode of transport, while in high-deprivation areas public transport (especially the bus) was the key mode of transport. In many societies, the car is a symbol of people's aspirations and social status. The public transit system may bear a social stigma, especially if it is perceived as dirty, unsafe, and unreliable. To fight such associations, the public transport services should be of highest possible quality to inspire a sense of pride in users, accompanied by awareness about doing good for the environment and the planet.

Public transport

High-quality public transport can reduce reliance on cars. The transportation network should give priority to public transit, pedestrians, and cyclists. The shift from a car to public transport may require people to start valuing the alternative differently (Boyko et al., 2021). People are willing to walk or cycle between activities if they see it as more convenient, healthier, cheaper, and more agreeable. People need to view individual car transport as a kind of failure mode and slightly toxic resource (Boyko et al., 2021).

The availability of public transport is one of the main conditions of successful eco-neighbourhoods. The location of new eco-neighbourhood needs to be carefully chosen to provide easy access to public transport. Stops should be located so that every resident can easily reach them on foot, and their path to the stop be safe and convenient. The transportation should be designed around walking and encounters with people (comfortable walking paths, places to sit and chat, coffee shops, small public spaces, public parks, etc.)

Pedestrian movement and walkability

Walkability is a concept associated with urban environments. It means that urban conditions of a given space promote walking. Those conditions are not limited to physical criteria (pedestrian paths, continuity of pavement, no obstacles) but also subjective criteria like comfort and relationship to the space (Alfonzo, 2005; Roussel, 2023).

The physical aspects of walkability are the domain of architects, urban designers, and planners. The sidewalk needs to be wide enough to accommodate casual activities, secondary functions (selling goods, restaurants' gardens, places to sit and chat). All of these activities translate into humanization of the city. They are critical to urban security (Kapoor, 2023).

Reducing the risk of crime improves the sense of security. Security can be supported at the architectural level by ensuring good visibility of pedestrian paths from common spaces and building windows. It is important to guarantee a sense of safety, which involves an even, non-slip surface, no obstacles, and places to sit where you can rest (every 200 m). Designing pedestrian-friendly streets involves limiting the speed and number of motor vehicles. Dispersing traffic by designing several alternative streets in a parallel system also helps to reduce traffic. The streets with less traffic are regarded by residents in a far more positive light. In areas with heavy traffic, people tend to move home more quickly and seldomly interacted with their neighbours. In such areas, those who suffer the most are the disadvantaged, children, and the old, as they are less mobile or able to move from home. Children living in high-rise buildings and dense urban environments are not likely to play outdoors (Boyko et al., 2021). Nevertheless, the street should be a space in the middle between the home and the outside world. It may be a place for children to safely explore (Biddulph, 2007).

Bicycle infrastructure

The connectivity of bicycle lines facilitates this active form of transportation. It is important to provide facilities for cyclists: paths, parking lots, service stations, bicycle shelters, and changing rooms with showers – next to public transport stops, next to workplaces, etc. The locker rooms with showers for cyclists located in public buildings, workplaces, and places of study may also encourage cycling. Service stations are also important, where people can quickly repair their bikes. The bicycle paths should be designed according to the same hierarchy of needs as pedestrian paths (Alfonzo, 2005).

References

Alfonzo M. 2005. To walk or not to walk? The hierarchy of walking needs. *Environment and Behavior*, 37(6), 808–836. https://doi.org/10.1177/0013916504274016.

Bartosz D. 2024. *Podręcznik Audytora Certyfikatu Zielony Dom v.2.01*. Polskie Stowarzyszenie Budownictwa Ekologicznego.

Biddulph, M. 2007. *Introduction to Residential Layout*. Architectural Press, Routledge, London. https://doi.org/10.4324/9780080468617.

Bonilauri E. 2015. The compactness ratio or form factor of a building, emupassive.com, published online on October 26, 2015, avaialble online: https://emupassive.com/2015/10/26/the-compactness-ratio-of-a-building/, retrieved on: 21.08.2024.

Boyko T., Cooper R., Dunn N. 2021. *Designing Future Cities for Wellbeing*. Routledge, London.

Declaration of Nyéléni. 27 February 2007. Nyéléni Village, Sélingué, Mali, available online: https://nyeleni.org/IMG/pdf/DeclNyeleni-en.pdf, retrieved on: 22.08.2024.

European Parliament. 2023. Circular economy: Definition, importance and benefits, published on 24.05.2023; last updated: 24.05.2023 – 11, available online: https://www.europarl.europa.eu/topics/en/article/20151201STO05603/circular-economy-definition-importance-and-benefits, retrieved on: 22.08.2024.

Friedman, A. 2022. *Designing Innovative Sustainable Neighborhoods* (1st ed.). Routledge, London. https://doi.org/10.4324/9781003203025.

IPHA International Passive House Association. 2021. What is a Passive House? Passipedia, available online: https://passipedia.org/basics/what_is_a_passive_house, retrieved on 21.08.2021.

Kapoor M.K. 2023. *Security by Design: Protecting Buildings and Public Places against Crime and Terror*. Routledge, London.

Lang J. 2017. *Urban Design*. Routledge, Abingdon, Oxon, New York.

Lawrence R.J. 2020. *Creating Built Environments: Bridging Knowledge and Practice Divides*. Routledge, Abingdon, Oxon, New York.

Kobylarczyk J. 2018. *Uwarunkowania środowiskowe w projektowaniu obszarów mieszkaniowych*. Wydawnictwo Politechniki Krakowskiej, Kraków.

Roussel J. 2023. Walkability. In: Marchand D. (Ed.), *100 Key Concepts in Environmental Psychology*. Routledge, Abington, Oxon, New York, p. 165.

Trojanowska M. 2020. *Poszukiwanie standard projektowania ekoosieldi w Polsce*. Wydawnictwa Naukowe Uniwersytetu Technologiczno-Przyrodniczego w Bydgoszczy, Bydgoszcz.

Wilson O. 1984. *Biophilia*. Harvard University Press, Cambridge, MA.

winnipeg.ca. 2024. 7R Basics, available online: https://guides.wpl.winnipeg.ca/greenchoices/basics, retrieved on: 22.08.2024.

World Bank. 2024. What a waste 2.0 a global snapshot of solid waste management to 2050, available online: https://datatopics.worldbank.org/what-a-waste/trends_in_solid_waste_management.html, retrieved on 22.08.2024.

Zielonko-Jung K., Marchwiński J. 2012. *Łączenie zaawansowanych I tradycyjnych technologii w architekturze proekologicznej*. Oficyna Wydawnicza Politechniki Warszawskiej, Warszawa.

5 Society & culture

Public space

Public spaces should contribute to public health. They affect how people feel about their community and place of living. The design and maintenance of public spaces, bolsters the social development and a sense of place. It is encouraged to promote variety to accommodate multiple uses. Open public spaces of various sizes can accommodate communal gatherings. Places which are busy should welcome more active, social forms of recreation while more tranquil spaces may be placed away from the main pathways (Friedman, 2022). Public spaces create casual environments and informal atmospheres for social encounters. These places have the potential to promote spontaneous interactions between people in the same neighbourhood.

Ray Oldenburg (1999) coined the concept of "third places" – third mostly frequented spaces after home and work/school, which are voluntarily chosen to visit. Patricia Simões Aelbrecht (2016) defined "fourth places" as informal settings alongside home, work, and "third places" which are inclusive public places "in-between". These are outdoor spots in which people can perform all activities that are in-between other necessary activities (stand or sit, people watch and interact) (Simões Aelbrecht, 2016). Thus, active sidewalks can nurture the social connectedness and trust between city residents or become alienating places (Boyko et al., 2021).

Public spaces can be divided into four categories: public, semi-public, semi-private, and private. A clear hierarchy and clear division of space contributes to improving the quality of life of residents. Public space is a space open to everyone (street, city square, public park). Semi-public space is a space open to everyone at certain hours and under certain rules (park open for a few hours during the day, internal courtyard in a housing estate). A semi-private space is a space that anyone can look into, but can enter only with the owner's invitation (private garden). Private space is a space to which only household members, family members, and people invited by them have access (private apartment).

The great problem of large housing estates was the lack of clear divisions of space. Anonymity is counteracted by a sense of territoriality, which can be supported by creating places that strengthen neighbourly ties: playgrounds for children,

DOI: 10.4324/9781003494898-6

meeting places for young people, recreation areas for the elderly, or community gardens. The introduction of such places also helps to build a sense of security.

No-man's space is space that no one cares about. Such spaces are often perceived as places that may be devastated without punishment. Public space is a public good and property of all of us in a visual sense. Contact with a devastated landscape reduces not only the sense of security, but also self-esteem. Beautiful spaces can boost self-esteem of users (Trojanowska, 2020).

Personalization of space has a positive effect on the subjective sense of security. Spaces such as separated and marked private gardens, community gardens, playgrounds, and parking spaces have a defined or presumed owner. Courtyards inside building blocks are spaces that belong to the residents of the quarter by default, which gives them a sense of ownership and protection against acts of vandalism.

We are all designers of public places: elected leaders, government officials, local people in business, public, cultural and leisure services, and local residents of all ages and social status (Boyko et al., 2021). We all can facilitate bringing people together. Jan Gehl (2011) discerns three categories of outdoor activities in the public areas: "necessary activities", "optional activities", and "social activities". Example of "necessary activities" is destination walking – participants have no choice about it. "Optional activities" requires the participants' will to engage (strolling, pleasure walking). "Social activities" are in contact with other people (children playing, friends talking, and greeting). In a well-designed physical environment, the optional and social activities occur with high frequency (Gehl, 2011).

Design for safety

Safety is the most important issue in public spaces design and maintenance. Even the most beautiful spaces would not be used if they are not perceived as safe. Lack of safety, social cohesion, and community involvement has a negative effect on child development (Minh et al., 2017).

Design for safety focuses on two fields: prevention and resilience. The first attempt is to prevent the incidence from happening in all realms: physical environment, socio-economic, cultural, psychological, etc. Ilan Kapoor (2023) defines the built environment design and construction quality as the immunity levels. When immunity fails and an incident happens, he declares the "resilience" as the second line of defence (Kapoor, 2023). Resilience is the ability to bounce back after any incident. It can be extended to ability to recover, adapt to change, and survive (keep going) (Ovans, 2015). Ilan Kapoor defines it with 3Rs for resilience – "Resist, Respond, and Recover". When protection is assessed to be insufficient or expensive, resilience allows for recovery. It is a fallback property of built environment to minimize losses from the failure of prevention strategies (Kapoor, 2023). The buildings can offer a second line of defence. They can operate in emergency mode. The structures can be designed to withstand the flooding, fire, seismic activity, or terrorist attack. The services may operate to allow people to evacuate.

Crime prevention through environmental design (CPTED) focuses on eliminating the opportunities for crime to be committed. There are some qualities of the environment which may deter possible crime. The primary goal of CPTED is to maximize visibility and foster positive social interactions. Lighting levels and surveillance by humans or cameras may affect the prevalence of incidences of crime. Uniformly distributed ample light levels may be boring, but they are the best for security (Kapoor, 2023).

The CPTED is used today to promote socially acceptable behaviour and to reduce the acts of vandalism and crime. Crime happens if there is an opportunity. If the opportunities are eliminated, the possibilities of crime would be diminished. The clear distinctions of space hierarchy – public and private identifies authorized users and helps to prevent crime. Ownership of a place signifies professional or casual surveillance, and that sends a subtle signal to the public. The mere presence of windows and balconies makes the looked-into space feel secure. There is a theory that a broken window invites more antisocial behaviour. The incidences of graffiti are regarded as an invitation to crime. On the other side, the signs of pride and ownership (Christmas lights, flags, and decorations) are a vital security tool. The maintenance and care are deterring crime (Kapoor, 2023).

It is important to provide a visual link between homes and neighbouring spaces and to give the inhabitants a sense of involvement with the public space. Street life enlivens places. Mixed-use development with commercial space on the ground and residences above contributes to safety and engaging environment (Friedman, 2022). The interesting and active frontages promote safety and security (Biddulph, 2007). Multifunctionality and mixed use of a building or neighbourhood may contribute to crime prevention. Functions that stimulate the presence of people during the day or night deter the isolation and discourage crime. Presence of people may have a positive effect in the form of "eyes on the street" with a brain to respond to a given situation. However, overcrowding may be a factor of crime opportunities. Too many people does not mean better surveillance (Kapoor, 2023).

The boundaries of public and private spaces are often marked by fences. Fences that mark property boundaries should not block the access to paths leading through the estate. The transparent, low fences and hedges that separate private gardens from walking paths work well. It is worth ensuring that the space is divided in a way that is user-friendly for residents, but at the same time clear in terms of conveying the message of proprietorship.

However, Ilan Kapoor (2023) draws the attention to the fact that opaque and impregnable boundaries give a false perception of security, as they disconnect the property completely from the larger urban context and renders it unsafe. A problematic issue may be the so-called gated communities. Fences are a psychological barrier and a problem for the accessibility of public spaces. The need to walk around the fenced estates discourages people from traveling on foot (Trojanowska, 2020). Gated communities develop auto-dependency. All non-residents are treated as potential perpetuators excluded from public areas (Biddulph, 2007). The security of gated communities is questionably. Unless they employ professional guards, they rely on the same natural surveillance mechanisms, which are based on reciprocity. If the neighbourhood is supposed to watch over a property, the property must contribute the same to the surroundings. Natural security is a two-way process of receiving and giving. Outside of the secured community the streets are fearsome (Kapoor, 2023). Security is the first casualty of moving from the familiarity of rural to the anonymity of the urbane. "The anonymity is detrimental to security" (Kapoor, 2023).

Good urban design and planning can stimulate the levels of interaction between the public, semi-public, semi-private, and private areas. Planning controls can be consistent within the city mandating open spaces and transparency of perimeter design. Walkability and humanization are crucial factors that promote safety and security. Traffic calming is a factor contributing to walkability, safety, and security. Where vehicles are mandated to slow down they contribute to natural surveillance.

The most fundamental need of residents is to ensure safety. At the design level, it is necessary to eliminate escape spaces or oppressive interiors. Both narrow and tight spaces, as well as empty open areas of land, arouse fear. It is worth ensuring good visibility with long viewing axes and the ability to observe other people's activities. Monitoring and possibly supervision by police patrols is also necessary. Neighbourhood surveillance and reporting of suspicious activity brings results, although it may also lead to acts of violence that go beyond the limits of self-defence.

People feel safe in clean and well-kept spaces. Maintaining cleanliness and order is important. Public spaces should be cleaned and waste bins emptied regularly. The sense of security increases in a clean environment. A visible lack of care (abandoned waste, overfilled waste bins) makes people feel uncomfortable in a given space. Expenditures on servicing and washing sidewalks, emptying garbage containers, and cleaning are necessary in eco-neighbourhoods. They can be limited if each resident feels responsible for the common space, but it is difficult to reduce them to zero.

The acceptance of changing seasons and natural processes of decay is a different problem. This means allowing the grass to turn yellow, leaving dry leaves and flowers in the winter. The sustainable methods of maintenance do not use artificial chemical fertilizers, and watering is limited. If those methods are used, the turf is not going to always look green. The new aesthetic of sustainable naturalistic landscaping encompasses all stages of biological life, thus yellow and brown turf as well. However, it requires acceptance on the side of users. Changing aesthetic beliefs may take time and require education.

Universal accessibility

Integrating the needs of people with special needs is a priority for friendly spaces. This is a very broad group, which includes, among others, children, elderly, pregnant women, mothers with small children, people with reduced mobility, people with intellectual disabilities, people after surgery, strangers, foreigners, people who don't speak the local language, etc. The eco-neighbourhood should be developed to ensure universal accessibility for the widest possible group of users. Temporary or permanent impairment of mobility cannot constitute a reason for limiting the use of public spaces. They should be designed to prevent accidents and to limit the risk of falling. The active sidewalks must be wide enough to allow two people in wheelchairs to pass one another. Clear paths should be unobstructed, level, and with a smooth surface. Design accessible ramps with low, accessible slopes at all crossings, and provide cut-through paths in medians, pedestrian refuge islands, and corner islands.

All users groups

Public green spaces are places for social connection, which is a key feature of places that promote health. Many health problems can be minimized thanks to physical activity or mental regeneration in a public park. People feel the need for social contacts and belonging to a group, but the need for contact with other people is conditioned by the psychological capabilities of a given individual, which change over time. Stigsdotter and Grahn (2002) develop a scheme of type of involvement depending on the individual's mental power. It can be translated into the pyramid of social contacts. At the very top is the leader's role (outgoing involvement), which requires the greatest vitality and is the fastest to exhaust a person's energy. Only people whose social networking abilities reach this level can start a project and see it through to completion, even without the support of others. Below is a level that requires slightly less involvement, but is also very absorbing – active participation in events requiring commitment and cooperation with others. The emotional participation level is the stage of regeneration when a person prefers to watch the activity of other people and animals, but does not want to take part in it personally. People at this stage are interested in other people, but this interest is passive, and they do not want to initiate contacts. At the base of the pyramid is the level of self-focus (directed inwards involvement). People at this level feel mentally exhausted and do not want to establish social contacts; they want to rest in solitude and silence. Self-focused people consider internal stimuli and physical activity that includes individual activities, e.g., walking or simple gardening. In the public spaces of the eco-neighbourhood, it is good to plan space for people's regeneration according to the levels of capability for social contacts in different periods of life (Stigsdotter & Grahn, 2002; Trojanowska, 2020).

All public spaces should be designed to accommodate various users' groups with different capabilities in a given moment of time. There should be places to rest and sit in quiet and solitude, observe the activities of other people from a distance (emotional participation), engage in active participation, and lead the others in various projects.

Henk Staats (2023) draw the attention to beneficial qualities of human-nature interaction resulting from relative absence of other people. The absence of continuous vigilance that is necessary in populated environments allows mental and physical regeneration. Public parks and gardens are offering space to stay in peace and quiet when it is needed.

Orientation

Mike Biddulph (2007) emphasized that the feeling of safety is connected to knowing where you are and where are you going. Orientation and wayfinding enhancement strategies help people feel safe within their urban environment. Humans need a clear pattern of main streets, with direct and obvious connections to other streets, to find way to their homes easily. However, a balance must be saved between creating safe environments and high-quality urban design. Kevin Lynch (1961) persuaded that built environments can help people navigate the space and fully identify with their surroundings. He lists five types of elements that create the content of the city images:

1 Paths
2 Edges
3 Districts
4 Nodes
5 Landmarks

Landmarks are creating a point of reference, and thus are useful for orientation. They are adding interest to urban landscape, often called townscape. Gordon Cullen (1961) describes planning a townscape a distinctive art. He argues that the landscape, topography, greenery, buildings, and space should create a distinctive urban character. They are giving a place the notion of placeness. Designing townscape should be a strive to achieve harmonious and beautiful scenography for human life. Mike Biddulph (2007) draws the attention to importance of designed views along a landscape corridor which offer enjoyment to people. People like openings to long compositional axes. Streets can be oriented so that they are closed by attractive building elements or landscape openings. Planting trees and shrubs or placing, e.g., sculptural compositions on the closed axis, may be helpful.

It is important to build an environment that facilitates spatial orientation and the creation of mental maps. The attractiveness of spatial forms in the surroundings also supports the development of a sense of local identity among residents.

Wayfinding

It is best if the entire city has a coherent Cities Information System (Visual Identity System), and a subsystem compatible with it has been developed for the eco-neighbourhood. This is a convenience for residents. It allows people to feel safer in an organized space. The developed scheme of signboards, signs, signposts, posters, and information placards should be compatible with the city system and the recommended patterns of safety signs.

Legible urban design

The quality of life is influenced by the functional and spatial structure and the way functions are arranged in space. The townscape quality is related to the surrounding landscape and the urban composition of the neighbourhood. Esther Sternberg (2010) pointed out that during difficult life situations – fatigue or illness, short-term memory capabilities are limited, and the ability to orient in space decreases. The neighbourhood should be designed to facilitate wayfinding. An orderly plan and a clear urban composition are very important. It is necessary to provide a sufficient number of characteristic features and details that distinguish a given space. It is wrong to allow the construction of repetitive building forms. The role of the urban designer is to find an appropriate, balanced form adapted to the natural landscape.

Architectural diversity

There is ample scientific evidence that the human mind requires a certain level of complexity organized in an orderly framework. Euclidean forms (circle, square, rectangle, triangle, cube, sphere) exist in the human environment, but they are only derivatives of human activity. The natural world that surrounds us is made of non-Euclidean forms. Benoît Mandelbrot created the foundations for fractal geometry, the mathematical language helpful in describing natural landscapes. The Mandelbrot set is one of the most famous concepts of modern mathematics, which allows for the description of complex forms of the world surrounding humans. Fractal forms found in nature create the infinite beauty of creation (mountain tops, tree crowns, forms of leaves and flowers, or a complex drawing of snowflakes) (Trojanowska, 2020).

People need environments with intricate details that stimulate the brain. Humans need multisensory stimulation, which translates into the need for diversity. An environment completely devoid of details, colours, and shapes can lower human's mood and even put people in a state similar to that experienced by those suffering from achromatopsia (Sternberg, 2010). Referring to the theory of biophilia, there is a need to design environments that imitate organically complex natural landscapes. The natural need to be in an inspiring environment is related to the preference for fractal forms and the golden ratio found in nature. Architectural diversity may have a positive impact on the well-being and health of residents. Exposure to fractal patterns helps speed recovery after surgery. There are many examples of referring to organic fractal forms in architecture (Gothic Cathedrals). Complicated patterns in architecture and design are usually based on complex fractal composition.

Solidarity

The development of eco-neighbourhoods is based on the activities of local governments, but the ecological awareness of citizens, and their concern for the common public good are of key importance. A sense of equality and acceptance is needed, especially in the context of health promotion. The concept of an eco-neighbourhood assumes co-residence in the spirit of social solidarity of people from different backgrounds, age, profession, property, with different levels of ability, etc. An atmosphere of social, inclusive acceptance of all, is important. A strong network of safe social contacts is the best way to prevent mental health problems. It is important to create a micro-society, i.e., a community with a profile diversified in every aspect (age, profession, income level). Diversity in age and income levels within the neighbourhood is intended to provide a better start for both children from poorer and more affluent families. This allows children to learn about different lifestyles and prevents the stigmatization of neighbourhoods that gather only the poorest families and people at risk of social exclusion.

The great problem of ghettoization of poor districts is the lack of role models. Solidarity and cohabitation of various social groups facilitates the development of social competences. Children have good examples around them. They learn by watching people achieve their goals, if the neighbours come from different social groups (white-collar, blue-collar, businessmen, freelancers, craftsmen, clerks, teachers, office workers, employees of various industries). Cohabitation allows for the reintegration of people at risk of social exclusion.

Creating an inclusive, accessible environment where everyone feels good promotes health. Today, well-developed social capital, both of a given individual and the entire community, is considered to be the most important factor protecting against life difficulties. A network of friends and kind people helps people to get through worse periods of life and return to healthy functioning in society. Research reveal the need for the context of "being with others" for the personal well-being (visiting family, meeting friends, going out, etc.). There are two different categories depending on their contribution to the well-being of an individual or the society. The first is active engagement with others that is promoting individual well-being. The second is purposeful engagement in society as a whole (attending community groups and social activities) (Boyko et al., 2021).

Public park is an inclusive, non-demanding place of spontaneous encounters and meetings. It may become the third or the fourth place. For many, regardless of social status, the public green space is a place for mental regeneration and physical activity. In the park adults choose more or less demanding forms of activity, depending on their level of physical fitness. Some may need paths that create loops for running, walking, or Nordic walking. For safety reasons, it is worth separating bicycle paths and the pedestrian path system. The more sports and recreational infrastructure is ready to be used in the park, the better: outdoor gyms, multipurpose sports fields, tennis courts, open green areas where classes can be organized (urban sketchers, aerobics, park run). It's good if there's a lot going on in the park; animating the place helps build bonds with the place of residence and the identity of the place. it is also worth considering building a swimming pool and a slide, an ice rink, or other facilities in line with the expectations of the residents (Trojanowska, 2017).

Restraining gentrification

The diversification of society helps to build social capital and resilience. The diversified housing programme and mixed-use neighbourhood may attract various social groups. Sometimes, there are incentives for the young offering good deals on first tiny flats. There are also initiatives to attract PhD students with suitable apartments for lease. Families with children would appreciate larger flats with balconies and a view of a playground. They want to have a public park and recreational facilities within walking distance. The elderly and retired would appreciate a smaller, affordable, and universally accessible apartment in a friendly neighbourhood with all necessary commerce and services. The coexistence of social housing and luxury accommodation is possible in mid-density and high-quality urban neighbourhoods. In eco-neighbourhoods in France (Éco-Quartier), where a system of subsidized apartments has been implemented, municipal social apartments with various levels of rent subsidies are offered, in addition to apartments for sale. It works well in very attractive location. There are also solutions involving the construction of modular apartments, which can be combined or divided into larger or smaller areas depending on current living needs – from studio apartment for single person to a few bedrooms apartment for multi-generational family.

Ageing in place

Accessible buildings and public spaces are helping all to age in place. The neighbourhood design may encourage seniors and other residents with special needs to stay active and engaged. Elderly people have less strength and stamina, thus they are less mobile and appreciate well-maintained wide active sidewalks with places to rest. Benches to sit and rest should be located every 200 m in public space to be more inviting to people with special needs. Safe paths, plazas, and urban green spaces encourage senior residents to spend time outside and build relationships with neighbours. Streets offer possibilities for physical activity, meeting people, and access to local amenities and facilities. Sometimes the seniors are using streets without any purpose, just to be outside and enjoy sunshine and fresh air, without any destination in mind. Poor street design and management condemns people to be trapped in their homes.

Good public transportation encourages seniors to move around the city. Mixed-use neighbourhoods with essential services decrease car dependency and enable seniors to live independently. Multi-generational housing helps people with special needs to interact with other people and increase the social capital. Living in close proximity creates intergenerational support system.

Design for aging in place requires adapting a barrier-free design to facilitate independent living for those who wish to remain in their current dwelling. New dwellings should be designed to allow for simple retrofitting (Friedman, 2022). The safety and security factors must be considered such as reducing stairs and thresholds, introducing elevators, even surfaces, seamless transitions, manoeuvring space for wheelchairs, space beneath the countertops in kitchens, accessible bathrooms and workplaces, and railings and handles to prevent falling. Adaptability of the dwelling to transformations for universal accessibility increases the social sustainability of the neighbourhood.

Design for the young

Children-friendly design is based on the simple fact that children will want to use the whole environment. Children will play in non-structured way anywhere in the neighbourhood. Designers must ensure that all public spaces are safe within reasonable limits. Children need safe walking and cycling paths to enjoy more liberty and explore the neighbourhoods by themselves, walking to school, local shops, and amenities. They like to play in residential areas, because they are easily accessible. However, the fear that a child would be hit by a car undermine the role of streets in child's play. Traffic calming within the vicinity of homes makes the environment a bit safer. Children are less capable than adults to judge the speed and assess the risks. Safe intersections for children should have highly visible pedestrian crossings, low traffic speeds, and signals timed for a slow walking speed.

Young people may be considered in three age categories: preschool children, young children aged 5–12 years, and teenagers (Biddulph, 2007).

Preschool children need space close to home. Tot lots should be located within 5 minutes from every home. Very young children require direct supervision by adults. Including spaces for preschool children to play under the gaze of parents is a challenge for the landscape designers. Those spaces should be a part of the green and blue infrastructure to provide enjoyment to all, whether they are used by children (morning) or adults only (evenings). Preschool children need to practice motor skills and enjoy water and sand.

Young children aged 5–12 years like to socialize with their peers and need to spend time outdoors for their development. Playgrounds may be equipped with colourful and expensive playsets, wooden custom-made facilities, or become an adventurous playground made by children themselves with help of animators and parents. Children are keen to be creative and experiment. They are drawn to creative play and eagerly manipulate their environment, build structures, and explore nature. Places of creative play are places of education and development. They stimulate children to discover truths about the natural world and challenge their ideas about their possibilities. It is worth emphasizing that creative play areas are created according to individual designs and do not have the repeatable forms of typical playgrounds from the catalogue. These are personalized spaces, designed in a specific location, adapted to the natural and landscape forms. They can become a focal point and activity nod in the neighbourhood and contribute to the creation of local identity (Trojanowska, 2020).

Teenagers do not want to play in the playground with younger children, and they also separate themselves from adults. At this age, young people feel a particularly strong need to belong to their peer group. They want to spend time together. Teenagers like to hang out in public spaces. They need seating and shelter either on visible street corners or in secluded locations, away from the gaze of adults. Public space which allows for neutral, active use is important for their physiological and psychological development (Biddulph, 2007). It is worth designing the space in line with the expectations of teenage users and including them in the participation process. For teenagers, you can design a sports and recreational space, including, for example, multi-purpose sports fields, cycling paths, and a skate park. Among favourite activities are cycling, skateboarding, and team games (soccer, football, basketball, etc.). Teenagers also need access to a wider range of more formal facilities like youth clubs, leisure centres, larger urban parks, malls, and cinemas, and therefore access to either cycle paths or bus routes becomes particularly important.

Children raised with electronic gadgets suffer from a deficit of both contact with nature and contacts with peers. Lack of contact with nature is sometimes described as a health problem of modern civilization (Louv, 2005). Vitamin G (G – Green) is necessary for the child's psychophysical balance. Children need shared play and a stimulating learning environment. This deficit is one of the factors contributing to many non-communicable diseases, obesity, and diabetes (Maas, 2008).

Basic needs

Pedestrian friendliness and high intensity are the two features most frequently mentioned in the literature on sustainable urban design in the context of health promotion. High intensity may be related to the proximity of public services, greenery, and workplaces. Local access to all primary services and commerces, healthcare, and primary education offers livelihood and the opportunity to meet all demands without the need to travel great distances for daily needs. Easy access to employment and education reduces the demands on transportation infrastructures and pollution.

Close access to basic services can improve the quality of life by eliminating the need to commute to destinations with car and finding parking spaces. Moreover, close access to services promotes active living and increases the physical activity of residents.

Building within the limits of existing urban fabric, and infill development, may involve additional costs, but it offers the wider selection of shared services, amenities, and proximity to public transport services that can significantly increase the market value. Moreover, public transport can significantly reduce the costs associated with using private cars.

Multifunctional neighbourhoods

Mixed-use development harnesses the virtues of secondary functions and casual security. It reduces urban stress, travel, and pollution (Kapoor, 2023). People adapt to new lifestyles quickly. Relocating to areas with higher neighbourhood accessibility to public transit decreases miles travelled in a vehicle (Friedman, 2022). Access to healthy food, childcare, healthcare, and other services fosters a sense of community, decreases vehicle dependence, and encourages residents to support local businesses (Friedman, 2022).

Each space can be multifunctional. Parking lot may be used for garage sales, farmers' markets, and large public gatherings (Friedman, 2022).

The concept of 15-minute city coined by Carlos Moreno in 2016 re-emerged during the COVID-19 pandemic (Moreno et al., 2021). He precises that the concept adds to the rhetoric of more humane urban fabric outlined by Christopher Alexander (1977, 2002). The concept of 15-minute city offers a novel perspective of "chrono-urbanism", specifying that all basic needs are located within 15 minutes' walk. The emergence of the COVID-19 revealed the vulnerabilities of car-oriented cities. The 15 minutes walkable neighbourhood caters for the needs of all age groups. It presents some form of social equity. Urban design enhances active transportation modes and promotes health. This concept requires the provision of all basic services, amenities, and facilities within walking distance of 15 minutes. As Friedman (2022) notices, to encourage inhabitants to use active mobility, there must exist a perception of achievability in the destinations they are aiming to reach. In situation where walking and biking are perceived as convenient and efficient, residents choose healthier option of active transportation mode. High density of people, places, and services saves energy and encourages people to walk. Accessibility of urban open green spaces promotes active living and in conjunction with walkability can improve the longevity of seniors (Takano et al., 2002).

Smart city

The development of smart city technology enables collective dynamic of actions. Along with the hedonistic purpose of good life, it also enables more eudemonic perspective of enabling the well-being of the greatest number of individuals. The smart city goals are improving urban efficiency and functional urban management (Boyko et al., 2021). Antoine Picon (2015) identifies two directions of smart city evolution towards support of health and well-being. The first is the provision of artificial intelligence for the management of the city and its infrastructures. Second is the coupling of stakeholders with government and infrastructure. The smart city facilitates monitoring of city functions and mitigating the risks of disaster. The eudemonic vision of good life is extended towards collective responsibility for planetary well-being beyond the individual life span and into future lives. The users are empowered to dynamically control themselves and their environment with new applications and services. The city is becoming a resource which can be shared over time to provide individual user benefits and collective gains (Boyko et al., 2021). The ubiquitous computing, augmented reality, and geolocation enhance every square meter of urban space and enable people, object, and spaces to tell their stories (Picon, 2015). The technology is forming a new skin of data across all the surfaces of the city forming an enacted space. The users gain the resources to author the city in a collaborative way. The virtual reality can be taken onto the streets. People are empowered to determine their conditions of living seeing the previously invisible digital structures, if the big data is free and open to use by the citizens.

Researchers emphasize the need to put citizens first in the approach to the smart city (Boyko et al., 2021). Open accessibility to digital processes makes a change in government interaction with people. The smart city data could empower networked citizens to innovate, share, and learn. It can strengthen the resilience of communities. The smart city technology can assert Lefebre's right to the city.

The self-determination empowered with smart cities data is fundamental for psychological well-being. The well-being is "politically enabled through technological empowerment" (Boyko et al., 2021).

However, there are some serious drawbacks resulting from availability of open big data. The children who are endangered with the lack of contact with nature and vitamin G are even less likely to play outside as they are overstimulated with digital media and technology (Boyko et al., 2021).

Local identity/place identity

The local identity is a difficult-to-define concept encompassing collective beliefs, myths, rituals, and non-monetary values (Lawrence, 2020). Place identity incorporates a shared view of a landscape perceived in a symbolic way. The process of attributing meaning to the landscape entails a potential for conflict between different social groups (Lima, 2023). Place identity links the memory with the city townscape (Jodelet, 2023). The natural space is linked to memory as a place marked by personal or collective history. The landscape may be marked by historic events (battlefields) and the townscape by architectural and urban forms (styles and historic periods). The built environment is reminding people of their past. Old buildings embody history.

Place identity is important for maintaining the continuity of memory and national identity. Monuments, commemorative plaques, and events commemorating important dates from the past are needed. Protection of cultural identity also requires preserving intangible cultural assets (historical geographical names, regional customs and costumes, traditional handicrafts). Cultural sustainability is the preservation of old and creation of new identities (Friedman, 2022).

The cultural landscapes require only minimal intervention to exhibit the background of the historic events. The concept of veracity is based on comprehensive documentation. "The diversity of cultures and heritage in our world is an irreplaceable source of spiritual and intellectual richness for all humankind" (The NARA Document on Authenticity, 1994). The places of memory are marked by three characteristics: identity, relational, and historic meaning. Jan Wrana (2011) defines a place as the result of the relationship between the physical attributes of a given space and the activities and concepts that people associate with it. He proposes recognize the "identification code" of the identity of the place in order to protect its "intangible content" while striving for a "new identity" of additions in contemporary form and material. He suggests preserving the existing with conscious conservation, modernization, adaptation, while maintaining the scale of the object, preserving and exposing valuable historical threads, and respecting the identity of the place. In case of introduction of a new feature, he recommends updating the record using creative continuation. Creative continuation respects the current context of the place in a new proposed form. A new context of the place is created by the contrast of form, sometimes treating the space as a material for artistic creation, using a pretext, a metaphor. Thus, creating new content and message.

There are many attempts to recreate local identity through architectural and urban planning. Such measures contribute to creating a sense of belonging among residents and attachment to the place. The way a place is perceived is greatly influenced by its inclusion in the city fabric. Clear boundaries and a marking of individuality are needed. This is especially important in the case of revitalized districts, whose prestige needs to be rebuilt. Entrances and gates to the estate may be marked with Visual Identity System (colour scheme). Taking care of the harmonious development of the neighbourhood should result in coherent aesthetics. The name of the neighbourhood is also important and should evoke positive associations.

However, the coexistence between communities that are distinct by their ethnic origins and by inequalities of resources may be source of problems (Jodelet, 2023).

Local culture

Culture contributes to well-being in very intrinsic ways. If the culture is demoted or regarded as secondary, passive, or ornamental support the contribution is discounted. Culture may be used as a tool to ease some identifiable problems or seen as something to be added when other issues are solved (Boyko et al., 2021). Urban inequalities are resulting in poorer health, quality of life, and well-being. They are a barrier to take advantage of the cities' offer of diverse possibilities for enhancing well-being. The cultural and social diversity are the sources of inclusivity, well-being, and urban health (Boyko et al., 2021).

The mixed-use neighbourhoods are offering cultural services (cultural centre, cinema, local theatre, art galleries). Organizing events is necessary to build the identity of a place, neighbourly bonds, and prevent social exclusion. The eco-neighbourhood needs places to organize a local event (family festival, a competition for children, or a local artistic event). What is needed is a place for a stage and comfortable stands for the audience. A local cultural centre may become the third place where both children and adults may want to spend most of their time after the first place – home and the second place – work or school. If it is not possible to build a separate building, the functions of a local cultural centre can be organized on the premises of a public school. The public park may become a space for cultural events (music concerts, open-air theatre performance, local fairs). Open clearings in public parks or grassy playing fields are often used. Spontaneous interactions during organized events foster creation of true sense of community. Gathering places allow neighbours to socialize together. The development of activity hubs in a mixed-use district contributes to a pleasant and healthy physical environment accessible to pedestrians, providing many opportunities for social interaction.

Spirituality and sacred space

The symbolic and spiritual dimension is important for every person. People need a place for spiritual renewal. Residents build chapels, place commemorative crosses, and figurines of Patron Saints. The path to the temple can be a place of spiritual renewal, which helps you forget about everyday problems for a moment and focus on timeless truths, reflection, and contemplation. Places with an established reputation as therapeutic landscapes, such as Lourdes in France or St. Anne de Beaupre in Quebec, Canada, combine the natural physical beauty of the landscape with sacred and symbolic features.

A person's spiritual needs are as important as material needs. The sacred dimension is important in the aspect of health promotion and cannot be neglected. Places of spiritual renewal such as churches, chapels, and sanctuaries often become the third place, chosen voluntarily, where people spend most time after home and office or school. Within the eco-neighbourhood, there should be places of worship and spiritual renewal within walking distance from apartments.

In Poland, during the communist era there were many grassroots bottom-up initiatives initiated by residents and paid for with great effort to build churches. They can testify to the striving of society to meet its spiritual needs. There were also accidents of social unrest facing the refusal of local or state authorities to build churches. Churches in Poland are funded by social contributions (crowdfunding).

The construction of a church in the ZAC Seguin-Rives de Seine à Boulogne eco-neighbourhood near Paris was also financed by private funds. In eco-neighbourhoods built in France, religious places, mainly churches or ecumenical chapels, are usually planned and built with financial resources from private funds.

Place attachment/placemaking

This term signifies the bond that individuals develop with a place. The affective and cognitive relation is based on positive emotions. Place attachment favours well-being and the comfort felt in these places. Individuals tend to protect places they are attached to and engage in pro-environmental (ecological, economic, and social) behaviours. People can experience a real connection with a place and invest meaningfully their space of life (Jodelet, 2023).

Placemaking covers a wide range of activities (public art, community events, and space animation). It is worth remembering that all activities that are intended to contribute to the creation of a friendly public space are also activities aimed at creating a place.

Mike Biddulph (2007) draws the attention to public art. It may be a valuable feature of residential area, but requires careful planning. Art used to reinforce a particular focal point may enhance urban design and reinforce local identity. Art may encourage a new meaning and sense of ownership. It could offer a cultural distinction to the area. Monuments acknowledge historic sites. Art most often takes the form of large-format installations (sculpture or murals) referring to local history, current events, or other content readable to the recipient. Art installations can also be a form of aestheticization of space, without additional educational or popularizing content.

Vernacular architecture

Architects have an influence on shaping cultural, mental, and environmental values. Today, the return to local identity is associated with a movement called critical regionalism. It is recommended to refer to regional construction by using forms, materials, details, and functional solutions typical of the culture of a given area.

Today we are witnessing the globalization of architectural designs. Regional architecture, which constitutes cultural heritage, is being replaced by buildings with forms far from the traditions of given regions. The forms of contemporary architecture, building materials, and details are repeated all over the world, regardless of climatic zones. Office buildings with facades made of system glass panes are a good example of globalized architecture. One of the main criticisms of modernist architecture was the duplication of cubic forms and materials such as concrete, glass, and steel throughout the world, even in regions where it was completely contrary not only to local building traditions, but even to common sense. Example is the irrationally high costs of transporting cement and steel to places far from production plants.

As an antidote, aspirations to preserve local identity, respect for the tradition and material culture of the place are being revived. Vernacular architecture, buildings built in specific climatic conditions, with local materials, age better, and do not require expensive repairs. Preserving local building traditions is also important to protect the skills of representatives of disappearing professions and crafts, e.g., carpenters. Traditional materials and techniques were chosen for specific reasons and are most often the most rational in many respects: ecological, economic, and social, i.e., rooted in the perception of the local community.

Architecture should be treated as an extension of the natural surroundings. Regionalism respects landscape and nature. We know many cases of shaping the building to keep landscape elements or living trees intact. Sometimes the entire location of the investment is changed to protect watercourses and natural habitats.

References

Alexander Ch. 1977. *A Pattern Language. Towns – Buildings – Construction.* Oxford University Press, New York.

Alexander Ch. 2002. *The Nature of Order: The Process of Creating Life*. The Centre for Environmental Structure, Berkeley, CA.

Biddulph M. 2007. *Introduction to Residential Layout*. Architectural Press, Routledge, London. https://doi.org/10.4324/9780080468617.

Boyko T., Cooper R., Dunn N. 2021. *Designing Future Cities for Wellbeing*. Routledge, London.

Cullen G. 1961. *The Concise Townscape*. Architectural Press, Routledge, Abingdon, Oxon, New York.

Friedman, A. 2022. *Designing Innovative Sustainable Neighborhoods* (1st ed.). Routledge, London. https://doi.org/10.4324/9781003203025.

Gehl J. 2011. *Life between Buildings: Using Public Space* (6th ed.). Island Press, Washington.

JodeletD. 2023. Memory of Places In: Marchand D. (Ed.), *100 Key Concepts in Environmental Psychology*. Routledge, Abington, Oxon, New York, p.80.

Kapoor M.K. 2023. *Security by Design: Protecting Buildings and Public Places Against Crime and Terror*. Routledge, London.

Lawrence R.J. 2020. *Creating Built Environments: Bridging Knowledge and Practice Divides*. Routledge, Abingdon, Oxon, New York.

Lima M-L., 2023. Landscape. In: Marchand D. (Ed.), *100 Key Concepts in Environmental Psychology*. Routledge, Abington, Oxon, New York, pp.75.

Louv R. 2005. *Last Child in the Woods: Saving Our Children from Nature-Deficit Disorder*, Algonquin Books, New York.

Lynch K. 1961. *The Image of the City*. The MIT Press, Cambridge, MA, London, England.

Maas J. 2008. *Vitamin G: Green Environments – Healthy Environments*. Nivel, Ultrecht.

Minh A, Muhajarine N, Janus M, Brownell M, Guhn M. 2017. A review of neighborhood effects and early child development: How, where, and for whom, do neighborhoods matter? *Health Place*, 46, 155–174. https://doi.org/10.1016/j.healthplace.2017.04.012. Epub 2017 May 18. PMID: 28528276.

Moreno C., Allam Z., Chabaud D., Gall C., Pratlong F. 2021. Introducing the '15-minute city': Sustainability, resilience and place identity in future post-pandemic cities. *Smart Cities*, 4(1), 93–111. https://doi.org/10.3390/smartcities4010006.

NARA Document on Authenticity, UNESCO. 1994. Available online: https://www.icomos.org/en/charters-and-texts/179-articles-en-francais/ressources/charters-and-standards/386-the-nara-document-on-authenticity-1994, retrieved on 28.08.2024.

Oldenburg R. 1999. *The Great Good Place*. Da Capo Press, Boston, MA.

Ovans, A. 2015. What resilience means and why it matters. *Harvard Business Review*.

Picon A. 2015. *Smart Cities: A Spatialised Intelligence*. Wiley, New York.

Simoes Aelbrecht P. 2016. 'Fourth places': The contemporary public settings for informal social interaction among strangers. *Journal of Urban Design*, 21(1), 124–152. https://doi.org/10.1080/13574809.2015.1106920.

Staats H. 2023. Nature, a psychological perspective. In: Marchand D. (Ed.), *100 Key Concepts in Environmental Psychology*. Routledge, Abington, Oxon, New York, pp. 87–90.

Sternberg E. 2010. *Healing Spaces. The Science of Place and Well-Being*, The Belknap Press of Harvard University Press, Cambridge, London. ISBN 978-0-674-05748-7.

Stigsdotter U.A., Grahn P. 2002. What makes a garden a healing garden? *Journal of Therapheutic Horticulture*, 13, 60–69.

Takano T., Nakamura K., Watanabe M. 2002. Urban residential environments and senior citizens' longevity in megacity areas: The importance of walkable green spacer. *Journal of Epidemiology Community Health*, 56, 913–918.

Trojanowska M. 2017. *Parki i ogrody terapeutyczne*, Wydawnictwo Naukowe PWN, Warszawa.

Trojanowska M. 2020. *Poszukiwanie standard projektowania ekoosieldi w Polsce*. Wydawnictwa Naukowe Uniwersytetu Technologiczno-Przyrodniczego w Bydgoszczy, Bydgoszcz.

Wrana J. 2011. Tożsamość iejsca kryterium w projektowaniu architektonicznym. Politechnika Lubelska, Lublin, available online: https://bc.pollub.pl/Content/646/PDF/tozsamosc.pdf, retrieved on 28.08.2024.

6 Case studies and best practices

Case studies and best practices – new capitals

There are quite a few new towns developed as the symbols and capitals of their countries or states: Washington, Canberra, New Delhi, Brasília, Chandigarh (1950s); Islamabad (1960s); Belmopan (1967); Dodoma, Abuja, Yamoussoukro (1980s); and Astana (1997). A few others are under construction: Putra Jaya in Malaysia, Naypyidaw in Myanmar (Burma), and Oyala in Equatorial Africa (Lang, 2017). Some other countries are also considering the relocation of their national capitals (South Korea, Argentina, Japan). Usually, renowned architectural practices are used to prepare plans for the development of new capitals. That was the case of the Chandigarh (Le Corbusier), Brasilia (Lucio Costa and Oscar Niemeyer), and Astana (Kisho Kurokawa).

Europe

Paris – 15 minutes city

Paris has an area of 100 km^2 divided into 20 districts (arrondissements) (Paris-city.fr, 2024). The city is a centre of region l'Île-de-France with an area of 12,012 km^2 and 12.1 million inhabitants. The climate is half-continental, half-oceanic with temperatures ranging from 2°C and 25°C. Annual precipitation is 600–700 mm (La prefecture…, 2022). The winters are cool, and the summers are mild to warm. Weather extremes are rare in this climate typical of Western Europe.

The 15-minute city concept "Ville du quart d'heure" was coined by Carlos Moreno, the French academic with Columbian roots. It was first presented at COP21 Conference in Paris (2015). The concept was derived from the works of Clarence Perry (1939) and Jane Jacobs (1961). It promotes mixed-use developments and compact, walkable communities. The concept is closed to principles of New Urbanism and Transit-Oriented Development. The idea is to create neighbourhoods where every day residents' needs are easily accessible in 15 minutes by walking, biking, or public transport ride. The daily necessities are basic services such as housing, work, education, shopping, healthcare, and leisure including accessible public green spaces. The concept filled with the pandemic and COVID-19 crisis

DOI: 10.4324/9781003494898-7

transformation of the established city life model. The 15-minute concept favours green and smart urban planning. This approach aims to promote health and sustainable lifestyle and reduce car dependency. It is connected with circular economy of short distances. To implement the 15-minute city some changes may be required involving multidisciplinary approach to polycentric city planning, transportation planning, urban design, and policymaking. Carlos Moreno emphasized the concept of "chrono-urbanism" focusing on value of time rather than cost (Moreno et al., 2021). The major components in the pursuit of 15-minute city are proximity, diversity, density, and ubiquity (Moreno et al., 2021). Carlos Moreno credits the work of Christopher Alexander (1977, 2002) and Nikos Salingaros (2000, 2006), who encouraged building more humane urban city fabrics.

The idea of 15-minute city was popularized by Ann Hidalgo, Mayor of Paris. She included plans to implement the concept with advices from Carlos Moreno himself. Other cities decided to follow the concept described as "a return to a local way of life" (O'Sullivan & Bliss, 2020; World Economic Forum, 2024). Among cities that implemented similar approaches in the wake after the COVID-19 pandemic were Copenhagen, Milan, Madrid, Melbourne, Ottawa, and Seattle.

Ecology

The major objective was to limit the perimeter of daily travel. When everything you need is located within 15-minute walking distance, people are less likely to use automobiles, thus saving on fuel, CO_2 emissions, and other forms of traffic pollution. Less automobile use requires less parking spaces. Decreasing number of onsite parking helps to increase the surface area of green space – allotments, playgrounds, and public parks. Larger surface of open green spaces translates to more biodiversity, less pollution, and better living conditions. The city strives to develop the interconnected complete grid of urban greenery – public parks, squares, pocket parks, and gardens linked by green streets with active tree canopies, green walls, and green roofs.

Economy

The 15-minute city concept promotes economy of short distances and circular economy. Numerous initiatives promote easy access to fresh food production, establishing cooperation with local food farmers and food producers. City dwellers can sponsor local farmers and commission a regular delivery of food baskets with seasonal production. Baskets can be delivered weekly or twice a month, depending on agreement. The agreement stipulates that the basket would be regularly delivered with a variety of fruits and vegetables. The content of the basket is always full of surprises – more of seasonal perishable production in spring, summer, and autumn and easy-to-store fruits and vegetables (apples, carrots, beetroots, onions, or potatoes) during the winter. The city of Paris encourages collective gardening and urban farming. Fashionable restaurants are offering seasoning with fresh herbs from urban gardens on the roofs.

The zero waste approach is also important for the economy. In Paris, the 7R zero waste principles are widely used and promoted: Refuse, Reduce, Reuse, Repair, Recycle, and Rot. Education about zero-waste lifestyle is offered to all age groups. Zero-waste shopping is encouraged. There are numerous zero-waste grocery shops. Food-sharing refrigerators are placed in available locations. Repairment workshops are organized in public buildings. Responsible fashion is promoted. Compost bins for organic compost production from plant waste are placed in all available locations – in front of public and residential buildings and in public parks and gardens. The ambition is to provide all of these amenities within 15-minute concept. Waste segregation is mandatory. In Paris, underground waste collection was introduced in renovated districts. Waste that could not be recycled is used as refuse-derived fuel (RDF) – in incineration plants for energy production. The waste management aim is to close the circle of circular economy.

Society

Jane Jacobs thought that a neighbourhood wasn't just about buildings, but social networks, too (Jacobs, 1961). Therefore, public buildings are used for multiple purposes, hosting different activities during workdays and at the weekends. The city of Paris aims to install local pride –"amour des lieux" – in urban dwellers. The development of local identity and social capital is promoted with various events. Walking for errands and for pleasure and spending more time in the neighbourhood offer more chances for establishing connections between people living in the same district. People are expected to spend more time together and know their neighbours better.

The development of open public green spaces is crucial for enhancing everyday contacts, small chats, and events. The municipality is encouraging establishment of collective gardens on public property lots. The associations of inhabitants are taking care of them. Two models are used. One requires assigning individual tiny lots to all members of the association. Every member is taking care of individual parcels. The other model is to establish a gardening calendar, and everyone can join to work on the entire parcel. The second model offers the joy of collective gardening to more people at the same time.

Human health

The major objective is to promote walking and cycling. Promoting physical activity is important for human health and well-being. Walking is a cheap and effective exercise for health-promotion. Tree-lined streets with wide pedestrian paths encourage walking. Cycling may be good for human health, too. The ambition of Ann Hidalgo, Paris Major, was to ensure that every street has a cycle lane by the end of 2024. Road security and reduction of accidents is envisaged. Less automobile use means reduction of urban pollution. Life quality has improved, thanks to easy access to services and public transport in multifunctional neighbourhoods.

Compact city is promoting physical and mental health. Time that is saved on everyday commute may be spent with family and friends or devoted to well-being and health. Increased leisure time may be good for health and can be dedicated to physical activity or favourite hobbies. Health services are accessible and available locally.

The concept of 15-minute city requires involvement of inhabitants in public life. Inclusive participation may lead to better mental health. Living in cities in general affects mental health. Advantages of living in the city, which may be good for mental health, include economic, cultural, and educational opportunities that you might not find in more rural areas. There are also numerous disadvantages. Mood and anxiety disorders are more prevalent in city dwellers (Lederborgen, 2011). Urban upbringing and city living have dissociable impacts on social evaluative stress processing in humans and may act as a risk factor for mental disorders (Costa E. Silva & Steffen, 2019). Various factors associated with city living may contribute to mental health problems: economic factors (unemployment, low socio-economic status), social condition (social network support), environmental exposures (toxins, air pollution, noise, light, sensory overload), urban design (contact with nature), and lifestyle (physical activity). The lifestyle offered by the compact 15-minute city – access to all the services and public transportation, physical activity and social capital, pleasant and safe neighbourhood, easy access to public parks, etc. – may ease the negative effect of city living.

ÉcoQuartier ZAC Clichy-Batignolles, Paris

The designers of this operation of urban renewal are architecte-urbaniste François Grether, landscape architect Jacquelin Osty and their team. The history of ZAC Clichy-Batignolles started with the ambition of the city of Paris to organize summer Olympic Games in 2008. This new urban district of Olympic village was planned to be a showcase model of a sustainable city planning. Eventually, the Olympic Games were organized by China, but the idea to promote the sustainability in new urban renewal project remained. The area of the project encompasses 54 hectares (ha) at the former SNCF rail yard in the north of the Batignolles neighbourhood (Paris 17th arrondissement). It is divided into three sections: ZAC Cardinet Chalabre (7.6 ha), ZAC Clichy-Batignolles (43.2 ha), Saussure (3 ha).

Ecology

This eco-neighbourhood is an example of urban sustainability. In the centre of the neighbourhood, a new park was constructed with an area of 10 ha. Martin Luther King Park is a haven for biodiversity – there are over 500 different species. The plants are omnipresent. Along with the Martin Luther King Park, green spaces were created inside housing blocks designed according to the Macrolot concept. Over 26,000 sq. m. of green roofs serve as ecological bridges, as well.

The park water needs are covered in 40% with rainwater, collected in a system of wet ditches and biotope basin. Only 12% of the district area is dedicated to driveways accessible to motorized vehicles.

The bioclimatic design of the buildings was paired with thermal insulation and elimination of thermal bridges. All the buildings have to meet the requirements equivalent to German Passivhaus standard for passive houses. The needs for energy for heating and cooling have to be lower than 15 kWh/sq. m./year. The demand for heating and hot water is covered in 85% with renewable geothermal energy. The electric energy is being produced on site with photovoltaic (PV) cells – 35,000 sq. m. – enough for satisfying 40% of the eco-neighbourhood electricity consumption (Clichy-Batignolles, Undated).

Economy

The design of this eco-neighbourhood limits the movement of private cars and reduces travel distances. The centrally located Martin Luther Park connects all of the parts of the neighbourhood, promoting walking and cycling to do every day errands. Inside the district 30 kmph zones promote soft trips. Accessible public transport was given a priority, and car parking areas were reduced. Therefore, the basic parking indicators are:

- 1 parking space per 100 m^2 of apartment space;
- 0.33 of 1 parking space per 100 m^2 of office space; and
- 0.28 of parking space per 100 m^2 of commercial space (Clichy-Batignolles, Undated).

To promote collective trips, new public transport hubs were constructed. One of the metro lines was extended, and new metro and bus stops were constructed.

Creating new local workplaces was another priority. The multi-use buildings offer commercial and office spaces on lower floors. The objective of 15-minute city – providing all the basic services – includes local employment opportunities. New office buildings were constructed in the Saussure section and along the railroads. Some of the largest corporations are renting office spaces here. The new Paris Courthouse designed by Renzo Piano Building Workshop accommodates up to 8,800 people a day (Paris Courthouse 2010–2017, Paris, France, 2024). Close to the ring road and railroad are located manufacturers and technical and logistic services (waste-sorting centre, concrete production plant, freight services, etc.). Businesses and retailers are encouraged to invest in ZAC Clichy-Batignolles and create new workplaces.

One of the technological advancement important for human health was AWCS Automatic Waste Collection Systems – vacuum underground waste collection. It is quiet and clean and reduces emissions by eliminating waste collection vehicles and compacting waste before disposal.

Society

The former rail area was planned to accommodate 7,500 of new inhabitants. Half of all the new apartments were planned to become social housing, 20% apartments with regulated rent, and only 30% were available to potential buyer.

The district was designed according to 15-minute city concept. Commercial, educational, and cultural centres and services are located within the 15-minute walking distance. The neighbourhood would be home to the new Cité du Théâtre, which would unite in one place – Odéon – Théâtre de l'Europe, new space for Comédie Française, French National Academy of Dramatic Arts, and accompanying services. A state-of-the-art cinema with seven-screen multiplex was built along with a new community centre. Inside the new eco-neighbourhood were established new schools, kindergartens, and healthcare services. Due to limited space they are located in multi-purpose buildings, sharing lots with other commercial, residential, and public services spaces.

The schools are offering equal opportunities to all students from the neighbourhood. The social capital creation is regarded as a priority. Numerous educational events for adults and children were organized during the construction phase. The city encourages local associations, who may use the premises of public buildings, schools, and community centre. It is envisaged to build strong community, who would promote local history and identity. The concept of social solidarity is emphasized with placemaking effort and inclusive design.

Human health

The Clichy-Batignolles neighbourhood was designed according to the Macrolot concept by Christian de Portzamparc. The concept is characterized by increased presence of green spaces and diversified architectural features, various heights, and fragmented perimeter. The major advantage of the Macrolot is sharing of the parking space, the facilities, and open green space (Guislain, 2016, 2019; Masbuongi, 2006). Each Macrolot could be built with approximately 10,000–30,000 sq. m. SHON. When individual green spaces are joined, it is possible to create larger public parks. This idea facilitated the creation of one of the largest public parks – Martin Luther King Park, which spans over 10 ha in the centre of the new district. Open green public spaces are accessible to all regardless of age and financial status. Both males and females could feel safe and welcomed. The functional and architectural diversity was promoted, each of the open blocks were designed by a different architectural office. The diversity encompasses mixed-use and various social groups of users and inhabitants. Walking is promoted. It is easier to run for everyday errands on foot or using a bicycle. The centrally located park is the heart of the district. All of the paths are crossing the park, which facilitates everyday contact with nature and biodiversity.

Healthy eating is crucial for human health. The grocery stores in the neighbourhood are offering seasonally available fresh fruits and vegetables. The inhabitants were invited to take part in community gardening in Martin Luther King Park. Workshops dedicated to healthy diet and cooking classes are organized by local associations.

Wrocław, Nowe Żerniki, Poland

Wrocław has an approximate area of 293 km^2 and is located at 130 m.a.s.l. (System Informacji Przestrzennej Wrocławia, 2024). Wroclaw has more than 674k inhabitants living in the city and 1,250 million people living in the metro area. Wroclaw has an oceanic climate (Cfb), bordering on a humid continental climate (Dfb) according to Köppen climate classification, with mean annual temperatures ranging between 0.5°C and 20°C (Feddema, 2005; visitwroclaw.eu, 2024; Oregon State University, 2024). Wrocław is the warmest of all cities in Poland.

Nowe Żerniki is a new district of Wroclaw developed to be a model and showroom of good practices in residential architecture. It was planned and designed to demonstrate alternative solutions to low-quality suburbanization and monotonous urban design. The district was developed as a joint effort of the city of Wroclaw, local branch of the Chamber of Polish Architects (DOIA IARP) and local branch of the Society of Polish Architects (SARP). Opposite to chaotic suburbanization around Wroclaw, Nowe Żerniki was planned as a new self-sufficient district with all public and commercial services needed for everyday living. The construction was planned for 2013–2023. However, due to economic and political reasons it is still under development. The district is planned to spread over 60 ha. The most important thing, in the opinion of the initiators of this action, was proving that for the same money you can create a better living space. The inspiration was the exhibition WuWA organized in 1929, which presented avant-garde architectural ideas for modern and comfortable living.

Ecology

The ecology was one of the important ideas of the new development. It is understood as furnishing the new buildings with newest technological advancements to preserve non-renewable resources and minimize costs of living. Integrated design was introduced from the very beginning of the process.

The neighbourhood is planned together with a public park located in the south of the new development. In the centre of the new district, there are new sport fields, playgrounds, and tennis courts. Numerous plants were planted in the neighbourhood, not only along the streets and avenues, but also in squares, plazas, and parks.

Nature-based solutions are used for rainwater infiltration. Part of the rainwater is collected and introduced into the water supply system as the so-called grey water for domestic purposes (irrigation, toilet flushing). That will reduce operating costs (Maćków, 2017).

Economy

The model of creating the neighbourhood itself is innovative. The owner of the land is one entity – the city of Wroclaw. The municipal investor made the area available to investors intending to build houses consistent with the master plan presented. It was important to involve various investors – developers, small cooperatives, as well as individual investors.

The design started with planning the appropriate scale and structure of the development. The neighbourhood was planned in such a way so that residents do not have to use a car. Walking for daily errands is promoted. Various solutions are planned to facilitate the use of bicycles – green paths, parking lots for bicycles, and bicycle rooms located on the ground floors of residential buildings.

The transportation needs were optimized to incorporate public transport whenever possible. All the basic services are offered in the neighbourhood, within walking distance. The tramway lines were extended, and the tram and rail stops were built next to the new neighbourhood, so that you can reach the city centre in a few minutes by public transport. Those who need may use it for daily travel to workplaces and educational services located away from the district, or for occasional travel.

One of the advantages of the master plan was diversity of the housing offer. Both large multi-family buildings and small units consisting of several apartments, as well as single-family terraced and detached houses were designed. Some of the buildings are planned using various forms of co-ops and public-private partnerships. The novelty is the implementation of model solutions for the development of buildings of various scales (Twardoch, 2018).

The economy of close distances and circularity was a priority. Local building materials and workmanship traditions were used. It was also important to create new jobs. Small co-working areas were proposed to invite local entrepreneurs to share office space.

Society

The new neighbourhood was planned to incorporate all urban facilities, such as school, kindergarten, nursery, church, clinic, home, community centre, marketplace, sports complexes, parks, squares, and commercial and service areas, which improve the quality of living. Urban design promotes walkability. In that sense, the new neighbourhood planning was close to the concept of 15-minute city. Public spaces were designed to serve residents as places of integration and recreation. There are solutions limiting car traffic.

The public spaces are aimed to promote social inclusion and integration. One of the major objectives was to guarantee security to all and prevent social exclusion. The aim of inclusive architectural design was to facilitate creation of social bonds. Care was taken to ensure the residents' sense of security, which results from belonging to the local community. Public participation was invited from the very beginning.

Therefore, the development system consists of small, clustered neighbourhood units within building blocks.

Streets, squares, and courtyards create a clear gradation of space. Private, semi-private, semi-public, and public spaces are distinguished. The intimate spaces between the buildings are used as semi-public spaces. Separated private space is secluded directly adjacent to the apartments.

Human health

Nowe Żerniki can be considered as an example of a modern garden city. A new park is planned next to the neighbourhood. The greenery is omnipresent – tree-lined streets, plazas, and courtyards. The design of the district puts people in the foreground. The development is organized into a series of quarters marking streets, parks, and public squares, designed for the pedestrians. Public space was designed as a place for interesting walks and meetings. From the beginning the "placemaking" was important (Maćków, 2017).

The innovativeness of the solution lies in its multifunctionality, inclusiveness, and human-friendly design. The human scale of public spaces, building blocks, and individual spaces was a priority. Architectural diversity resulted from the fact that each block was designed by a different architectural office. Spaces for living, work, and entertainment coexist.

North America

The cities in North America are undertaking numerous initiatives to counteract against destruction of natural environments and protect the health and well-being of its residents.

The city of Ithaca may serve as an example of good practices.

City of Ithaca, NY

Ithaca, NY, has a population of 31.4k people and area of 15.75 km^2 (DataUSA, 2024). The city is located at 123 m.a.s.l. City of Ithaca has a hemi-boreal, humid continental climate (Dfb) according to Köppen climate classification, with average annual temperature of 20°C and mean temperatures ranging between 8°C and 29°C (Feddema, 2005; weatherandclimate.co.uk, 2024b).

Most people in Ithaca, NY, walked to work, and the average commute time was 18.4 minutes (DataUSA, 2024). The city of Ithaca is home to Cornell University. The most common employment sectors for those who live in Ithaca, NY, are educational services (43.8%).

The city has taken many steps to become a more sustainable organization and promote sustainability and climate protection throughout the community. The priority is to save the planet while supporting a diverse, equitable, and vibrant economy.

Ecology

The city of Ithaca has been named a Tree City USA for 35 years. The Forestry Programme is recognized as a solution to create a cleaner and safer environment for residents. The Urban Forestry Programme adopted by the city authorities stipulates that Ithaca's community forest "should be multi-aged, fully stocked, healthy, and safe". The municipality maintains a full inventory of all trees to allow the city to focus on tree diversity and appropriate planting practices.

The city generates its own topsoil and produces its own mulch. The excess waste wood products are sold to a company which uses the waste wood as a biofuel for its production.

The policy of green purchasing is a good example of sustainable practices, too. All the cleaning and sanitary products that are purchased for distribution to all city departments should be environmentally friendly. Primary office product supplier provides recycled and other environment-friendly alternatives of basic materials (paper, ink cartridges, etc.) to public offices.

The city identifies potential, achievable steps the city can take to reach its energy and carbon reduction goals. The city is planning to reduce greenhouse gas (GHG) emissions from all local government operations by about 14%, compared to its 2001 baseline. Proposed actions include areas such as energy and fuel efficiency, renewable energy, land use and transportation planning, improved record-keeping, and other strategies (City of Ithaca, 2024a, 2024b, 2024c).

Economy

Ithaca Farmer's Market, created in 1973, helps promote local agriculture, thus strengthening the local economy. The market reduces the food transportation costs and carbon emissions associated with non-local food products.

Basic measures to reduce energy consumption by office appliances are encouraged in all public buildings: lighting upgrades and controls, facility management systems and other automatic temperature controls, building envelope improvements, high-efficiency furnaces and condensers, window replacements, steam traps, and more.

The municipality organized Sustainable Energy Education programme, which engaged all city employees and provided training in different subject areas and how we can make a difference through our own behaviour (i.e., turning off all devices at the end of the day or when leaving for extended periods).

The city is offsetting city government emissions by purchasing Green-e Energy-certified renewable energy credits (RECs) for 100% of its electricity, starting from 2012.

The Ithaca Youth Bureau is using geothermal heating and cooling system. In winter, the system absorbs the heat from the earth and distributes it throughout the building, and in summer it absorbs heat from the building and returns it to the earth.

Solar thermal hot water systems were installed on public buildings.

The city has a Green Fleet Policy and is using various strategies, such as the promotion of alternative means of transportation, reduction of vehicle miles travelled, improved fleet maintenance, purchase of high efficiency vehicles and alternative fuels, and educational programmes for employees.

The city of Ithaca has a very high pedestrian commuter percentage – over 40% of commuters being pedestrians. The increased pedestrian traffic is regarded as a factor improving the downtown economy, as people walking past local shops are likely to patron them. Therefore, the city is implementing pedestrian improvement projects, i.e., new sidewalk construction, intersection safety improvements, traffic signal upgrades, and snow-clearing requirements for sidewalks.

The city installed "Shared Roadway" signs and bike lanes to increase the opportunities for bicyclist. Bicycle parking added to businesses, apartments, schools, offices, etc. must meet high standards of a new Bike Parking Ordinance. There are routes where bicycle traffic is prioritized over motor vehicle traffic.

The city is purchasing biodiesel to use on its diesel-fuelled vehicles and equipment.

In addition to these initiatives, the city also provides free bus passes to city employees.

All city-owned traffic lights were replaced with LED bulbs.

The city is using Warm-Mix Asphalt technology (WMA), called Low-Emission Asphalt (LEA) (City of Ithaca, 2024a, 2024b, 2024c).

Society

The city is encouraging community and social bonds through series of action. Education about sustainable and healthy living and various programmes are offered. The Ithaca Youth Bureau Recreation Department offers year-round recreation programming for youth and families. Sports, lessons, camps, expressive arts, theatre classes, and more are offered. All who are interested may enrol in programmes online.

Human health

The Urban Forest Programme promotes creating healthy air, providing a cooling effect, and maintaining habitats for a variety of animals.

The city of Ithaca advocates for walking, instead of using a vehicle in order to promote improved physical health and reduce air pollution.

Healthy active lifestyle in the Ithaca community is promoted with the construction of scenic trails and promotion of non-motorized forms of transportation and recreation, walking, and bicycling.

North Downtown Athens Development, GA, USA

Athens-Clarke County, GA, has a population of 216,194k and area of 306 km^2 (DataUSA, 2024).

Athens has mild and moderate climate, humid subtropical climate (Cfa), according to the Köppen-Geiger classification. The average temperature in Athens is 16.9°C, with mean annual temperatures ranging between 6.3°C and 26.3°C. Average annual precipitation is 1231 mm. In Athens, there is a lot of rain even in the driest month (Climate Data, 2024a; Feddema, 2005).

The formerly Bethel Midtown Village and College & Hoyt senior community are planned to become a vibrant residential community at the northern edge of Downtown Athens, Georgia. The project – North Downtown Athens Master Planning Study – was awarded for Outstanding Public Involvement. The award is given in recognition of an initiative which involves the general public in a planning activity (GPA, 2024). The development received the LEED for Communities certificate.

Ecology

Building massing, orientation, programming, and materials selections were carefully checked to provide the best solar orientation and comfort. Placement of windows, balconies, and wall finishes were verified, as they can have an effect in energy performance and occupant comfort radiant temperature. The urban design was optimized for indoor and optimum outdoor thermal and visual comfort. The factors taken into consideration included solar radiation, relative humidity, and wind speed. The objective was to eliminate the heat stress or cold stress felt by the human body (JHP, 2021; JHP Architecture/Urban Design, 2023).

The storm water retention would be integrated into streets design. The wetland park is planned in the area with excessive run-off prone to flood risk.

Green interior courtyards are planned to become part of green and blue infrastructure.

Economy

The aim of the project is to connect the new development to public transportation and public green spaces and the activity of downtown Athens. The integrated design was incorporated from the very beginning.

All blocks of buildings would be multifunctional, with active frontages with services and commerce.

The infrastructure should be planned for future adaptability to renewable energy (North Downtown Athens Redevelopment, 2024).

Society

The priority for this project was the community outreach and involvement. The master plan design was based on a year of community outreach. The aim was to create a community that truly represents the aspirations and needs of the local residents, broader city community, stakeholders, and local leaders.

The objective of the project is to preserve and expand downtown affordable housing.

One of the important points mentioned by the community members was to reintroduce the historic city street grid and highlight the history of the neighbourhood. Art and history should be integrated into the new development.

Human health

The master plan presents a vision for a healthy, sustainable, and affordable community. Multiple parks and open spaces are planned throughout the community. The green trails are planned for safety and accessibility.

Everyday contact with nature is promoted with green streets, public spaces, plazas, linear park, wetland park, community park, and rain gardens. In the green interior courtyards gathering spaces, active spaces and community gardens are planned. The streets will be planted with shade trees. Adult exercise stations, playgrounds, and tot lots will be constructed to provide opportunities for physical activity (JHP, 2021; JHP Architecture/Urban Design, 2023).

Having influence on planning and decision-making is an empowering experience for members of local community.

MZT Aerospace Park in Mexico

MZT Aerospace Park is located in Sinaloa County. Sinaloa has a couple of climates: Mediterranean, hot summer climate (Csa), hot semi-arid climate, subtropical steppe (Bsh), and hot desert climate, subtropical desert (BWh), according to the Köppen-Geiger classification. The yearly average temperature is around 28°C, and the annual precipitation is 70 mm (Weather and Climate, Sinaloa, 2024).

Mazatlán weather is hot semi-arid climate, subtropical steppe (Bsh).

The temperature is approximately 24.3°C, and the rainfall is 722 mm (Climate Data, 2024d).

Mazatlán has a population of over 500k and an area of 3,068.48 km^2. It is located at an elevation of 10 m.a.s.l. (Gobierno de Mexico, 2024).

In 2020, 41.9% of the population in Mazatlán used own vehicle (car, truck, or motorcycle) as the main means of transportation to work (Gobierno de Mexico, 2024).

This is an example of sustainable proposal for industrial park – a research and development hub for aerospace industry in Mexico, Sinaloa district, city of Mazatlán (Industrial Park, 2024).

It will be the first Green Industrial Community in Latin America with LEED certification endorsed by the USGBC in Washington. Potential footprint is 203 ha. This new development is envisaged to stimulate the economic growth of the city of Mazatlán and region of Sinaloa. The new development is an example of good practices because the sustainability and well-being of employees were taken into account (El sol de mazatlan, 2021).

Ecology

The sustainability was an important point for this development. The park would use clean electricity, natural gas service, and state-of-the-art telecommunications transported by fibre optics and satellite. It is planned to re-use 85% of the treated water for the irrigation of the abundant green areas (MZT Aerospace Industrial Park, 2024).

The LED streetlights are used for illumination.

The ecological hydraulic concrete is used for avenues.

Economy

The MZT AeroSpace Park is part of the TMEC Corridor, a logistics megaproject which would take 15 years to complete (Tibesa Bienes Raices, 2024). It is built from scratch on a previously undeveloped land with all the necessary infrastructure.

The ambition is to stimulate the growth of the aerospace industry in Mexico, as it would be the only aerospace industrial park in the country. It is planned to be a state of the art, ultra-modern space for research and development, and production of new models of airplanes, drones, etc. The Industrial Park would have its own private runway. That would allow traditional aircraft to take off as well as experimental flights to carry out their flight tests. The investment is expected to bring a large boost to the economy and gain significant profits.

The development of the MZT AeroSpace Park is expected to stimulate the economic growth of the entire region. The envisaged investment of 330 million dollars should create 65,000 new jobs – directly on site and indirectly in the entire country (MZT Aerospace Industrial Park, 2024).

Society

The social welfare is an important part of the investment. The development would allow the smart business ecosystem to take advantage of human resources. The Universities in Mazatlán are offering various graduate programmes in engineering and technology; however, the region was plagued by the talent drain. The major advantage of the project would be the creation of new job opportunities for young generation – jobs linked to the technology sector, as well as for technicians of various vocations: electricians, mechanics, welders, and personnel who can work the machining of parts – with better salaries (Grave, 2021).

The city is collaborating with a German group that is highly specialized in technical training. Moreover, the MZT Aerospace Park would create opportunities for new professions and specializations in highly skilled engineering sectors. The city would also like to rescue lost trades, for example, cabinetmakers (Grave, 2021).

On the other hand, the city of Mazatlán is a famous tourist destination and would strive to maintain that reputation.

The MZT Aerospace Park would host not only businesses, but also facilities important for the well-being of employees and their families, like residential areas with recreational infrastructure, and kindergartens.

Human health

The aspect of health promotion in MZT Aerospace Park is based on providing sport and recreational infrastructure: two soccer fields, two multipurpose courts, outdoor gym, and walking areas. Moreover, some means of sound protection would have to be employed close to the aircraft runway.

South America

Latin America is still struggling with processes that are triggered by housing deficit, lack of social infrastructure, and by the need to improve the life quality of people that still live in poverty. Therefore, the examples of good practices are needed.

Curitiba, Brazil

Curitiba has an oceanic climate (Cfb) according to the Köppen-Geiger classification (Kottek et al., 2006). Average yearly temperature is 17.2°C with mean temperatures ranging between 20.6°C in summer and 13.4°C in winter. The precipitation level on a yearly basis is 1,630 mm (Climate Data, 2024c).

Curitiba has a population of over 1.8 million inhabitants and area of over 432 km^2. It is located at an elevation of 930 m.a.s.l.

The city of Curitiba, located in Parana State, Brazil, is considered a pioneer in sustainable city growth and quality urban planning. The vision for this approach started in 1968, when the city adapted Master Plan for Curitiba – well before the introduction of the word sustainability (Hidalgo, 2014).

The redevelopment and expansion of the existing city of Curitiba applied the principles of the functional city, with recognition of limited availability of public funds for infrastructure and restrained social access to the automobile (Jenkins & Smith, 2024).

Nowadays, the city population is over 1.8 million, and even though there are many problems the municipality is facing, the city is recognized internationally for its exemplary transportation system, progressive social services, and environmental practices. Among the factors which facilitated the city success are innovation, pragmatism, political leadership, and continuity. However, the city is facing similar ubiquitous problems as are other cities in the world: limited public funding, overcrowding, poverty, pollution, and flooding. Moreover, the city discourse as the world leader in sustainability may be hindering its future progress (Martinez et al., 2016). Curitiba's role as a model of sustainable city becomes so powerful and widely accepted that the city finds it difficult to live up to this standard. Still, the innovative practices in urban planning and management can be an interesting lesson for other cities around the world.

Ecology

The city is well-known for its environmental policies. Back in the 1980s the city created a number of urban parks, dedicated to various ethnic and immigrant groups. The policy of extending the city green and blue infrastructure is continued till nowadays. Curitiba has the highest rate of green spaces in Brazil and in the world. The greenbelt around the city was cultivated and preserved from urbanization. The city has developed parks in post-industrial areas and in floodplains. Parks have been extensively planted with trees.

One of the major challenges connected to resilience in the city of Curitiba is flood prevention. There are many rivers that cross the city, and the city has mitigated flooding through the strategic planning of urban lakes and retention ponds located in open green spaces. The lakes and ponds are functioning as water-flow regulators during the rainy season. Most of Curitiba's parks, called linear parks, are located along riverbanks and valley bottoms. In some neighbourhoods innovative solutions were needed (nature-based solutions, vertical gardens) (Hidalgo, 2014; Cabral, 2022).

The honey gardens programme was introduced to educate the inhabitants about the virtues of the native stingless bees as pollinators. Numerous beehives were placed around the city, in public community gardens, schools, etc. The efforts were multiplied to create a welcoming habitat and enlarge the population of native bees in the city (Lopes, 2023).

Economy

Curitiba is the country's fifth largest economic hub (UN, 2024). The most outstanding thing about Curitiba sustainable practices is the fact that its planners and managers implemented some creative and inexpensive ways to solve the universal problems of cities.

The urban design of the city is based on the trinitary system, which integrates mass transit, access roads, and land uses together. To implement that idea a progressive public transportation system was introduced in the 1970s – bus rapid transit (BRT). It is an integrated transport system based on exclusive bus lines extending throughout the metropolitan region. That inexpensive and efficient system was later introduced in many other places around the globe and currently operates in 168 cities (Global BRT Data, 2024). The system is one of many initiatives to fight the unemployment and poverty. The ticket price is the same regardless of the distance.

The roads are arterial in a way that there are wide lanes – bus passes for the 2–3 length buses, which are capable of transporting up to 300 people each. Today there is a BRT line: Linha verde – running on biofuel along a 22-km-long route (AFD, Undated; Prestes et al., 2022).

The city was successfully attracting new industries (automobile companies, new technologies). For the lower-income citizens, the municipality introduced job training and small-business incubators. The city established the Curitiba Industrial City (CIC) on the city's west side. The polluting industries are not allowed. The creation of CIC created over 200,000 direct and indirect jobs.

The city ambition was to preserve historical areas, revitalize neighbourhoods, construct new cultural sites, and attract tourism. The tourism sector is increasingly important to the local economy. Throughout the city the municipality has implemented several pedestrian-only and bike-only routes. Today, the city is perceived as a must-visit destination for all "intelligent travellers" interested in sustainability at both the global and local levels, an example of a city "doing it right" (National Geographic Education Blog, 2007).

Society

The municipality is striving to offer the inhabitants good quality of life. The city is taking some active steps to fight the poverty. One of them is *cambio verde* – green exchange is a programme which fights poverty, food insecurity, and food waste, at the same time as it is promoting recycling and healthy nutrition. The programme is based on exchanging recyclable materials (metal and glass) for fresh food (vegetables, fruits) – 4 kg of materials for 1 kg of fresh produce. The fresh food is produced by local farmers who live around Curitiba green belt. The city pays them a fair price for their production. It is a strategy where everybody wins. The inhabitants are getting fresh agricultural produce from "farm to table" and are also encouraged to recycle. The programme is available to all, including children and the less privileged. Thanks to these progressive practices, Curitiba is recycling around 70% of its waste (Cabral, 2022).

The other innovative solution to poverty is the programme of Linha da Economica. The old buses are converted into shops with government-subsidized basic goods. The programme is also subsidizing the local farmers by selling blemished or overrun produce, which otherwise might have been destroyed (Gnatek, 2003).

A public housing programme helps the urban poor to build their own houses.

The municipal university of sustainability Universidade Livre do Meio Ambiente is offering free education for future leaders in sustainability.

One of the innovative solutions for placemaking was the "citizenship street" – mini malls of government branch offices located throughout the city at major bus terminals.

Human health

Various programmes were introduced to educate the citizens to use their car less, recycle materials, and care for a healthy diet. Curitiba has been successful in reducing traffic congestion and pollution.

There are a few initiatives to promote health nutrition. Cambio verde was one of them. Another one is the Urban Agriculture Programme. The objective was to promote access to quality food, and reduce food insecurity, as well as promote greater social equity, support climate justice, and reinforce the city's green and blue infrastructure with natural crops. The programme promotes the use of public and private urban empty spaces, mostly under power grids. Urban farms are offering free guided tours, courses, training, and workshops to the community in general. The community associations are receiving help in soil preparation, technical guidance, seedlings, seeds, and other basic supplies. The Urban Farm is equipped with solar and wind energy generation systems, as well as rainwater harvesting systems. The vegetable gardens are built with recycled materials. There is usually a greenhouse and a composting facility on each site (FCI, 2024).

The programme called "leisure bus" Linha de Lazer is offering games, health education, and various art activities to children who do not have physical education curriculum, the elderly, and poor inhabitants (Gnatek, 2003).

Curitiba is one of the greenest cities in the world – 20% of the city surface are green spaces, approximately 52 sq. m. per person. Public parks were established in post-industrial areas and former quarries.

Bogotá, Colombia

Bogotá has an approximate area of 655 km^2 and more than 7 million inhabitants (www.dane.gov.co). The city is located at 2,640 m.a.s.l. and has a temperate warm climate, with mean annual temperatures ranging between 7°C and 18°C (Dobbs et al., 2018).

In 2022, Bogotá was among the 100 cities in the world that have stood out in 2022 for their advances in sustainability, analysing the factors framed within the Sustainable Cities Index (SCI) (Reyes, 2022). However, back in 1990s, Bogotá, the capital of Colombia, was one of the world's most dangerous places, associated more with drug wars and murder than sustainability. Everything changed in 1998, when Enrique Peñalosa was appointed as mayor. Even though the city is still among Latin America's most unequal cities, Bogotá may be regarded as a true source of inspiration.

Ecology

The city tried to incorporate into urban planning the management of ecosystem services.

Even though there is a wide range of green spaces Bogotá is hardly a green city. Even including the forest reserves in Bogotá's eastern hills, Bogotá residents enjoy just 5 sq. m. of green space per capita. Moreover, in general, the poorer the neighbourhood, the less green space there is. It is estimated that that approximately 40% of Bogotá's urban areas had an informal origin (Bernal et al., 2022). The informal settlements are usually built in fragile ecological areas. The self-built incremental housing in poor, informal neighbourhoods lacks water and sanitation services, as the municipality treats less than 30% of its wastewater. Most of informal neighbourhoods in Bogotá had been legalized, but spatial characteristics, morphology, and location, as well as an unequal public investment in infrastructure, remain problematic. Poor neighbourhoods face numerous threats including flood and landslides. Food insecurity is also a problem. In 2000, the Programme for Neighbourhood Integral Improvement, PMIB (Programa de Mejoramiento Integral de Barrios) started. This includes comprehensive intervention in public spaces, social facilities, and mobility infrastructure of informal settlements. Despite all efforts, spatial inequality continues. In Bogotá, statistics about access to environmental assets, adequate public spaces, and healthy environments still reflect a huge gap between the more affluent areas of the city and the poorest (Bernal et al., 2022). An equitable distribution of green infrastructure could improve the quality of living with access to public space and nature. Over the last 30 years the city has improved its environmental condition (Dobbs et al., 2018). Therefore, urban agriculture emerges as a multifaceted remedy for production, sustainability, and well-being. In Bogotá, residents have designed, self-built, and managed public spaces and community gardens for food production (Bernal et al., 2022).

Ecobarrios

First, ecobarrios in Colombia were established in Cali in 2018 and Bogotá in 2021. This initiative is supported by the French government (founder of Ecoquartiér movement in France) (Francia reafirma…). The objective of the programme

Ecobarrios is to reorient Bogotá's urban planning and construction actions, to be focused on sustainable development. The programme promotes sustainable practices in neighbourhoods and individual homes.

The neighbourhood La Perseverancia was the first eco-neighbourhood in Bogotá. In the district the system for storm water collection and reuse was implemented. The rainwater is filling the automated drip irrigation system, operated with PV energy.

In the case of El Cortijo-Ciudadela Colsubsidio, another eco-neighbourhood in Bogotá, sustainable practices include energy savings, use of renewable energy sources (PV), and recycling of rainwater for non-potable use. A strive for biodiversity protection led to installation of gardens for pollinators, bird houses, and feeders for house pets (Ramirez, 2022). In 2023, there were four eco-neighbourhoods in Bogotá and two others were planned (Barrero, 2023).

Economy

Bogotá was awarded the Sustainable Transport Award two times:

- 2005: for its TransMilenio BRT System (BRT) and the city's commitment to integrating bicycle infrastructure with mass transit, and redefining and reclaiming public space for its citizens (Sustainable Transport Award, 2005).
- 2022: for transport interventions that expand safety and mobility for the city's most vulnerable residents.

Through the 1990s and early 2000s, the city's authorities introduced a BRT system and developed more than 300 km of cycle lanes. BRT system plays a significant role in social integration of the polarized city – the poor live in the south and the rich in the north. BRT is not just a form of transport. It is a chance for a better life given to the poor (ITDP, 2009; Jara-Moreno, 2010). Youth can access better education, receive a university diploma, and find employment (Brodzinsky, 2014). There are also other benefits. The implementation of BRT resulted in reduction of car accidents.

TransMilenio was the only large-scale transportation project approved by the United Nations to generate and sell carbon credits (Brodzinsky, 2014). The city has been successful in lowering the air pollution and management of GHG emissions.

The use of bicycles is promoted with various policies, like car-free zones on Sunday.

Consumption of energy in Bogotá increasingly influences the development, economic growth, and welfare of the city population. In the city, the transportation sector is the greatest consumer of energy. The service sector has increased its share of energy consumption over the last decade, whereas the industrial sector has reduced its relative consumption (Martinez, 2015). The energy technologies and distribution systems are the main factors that shape the city and determine the needs and the resilience of every settlement. The municipality is struggling to lower the energy consumption and use renewable energy. The major source of energy for the city is generated by water power plant. To lower the transportation demand, the city incorporates electric buses. The city of Bogotá is among the three largest e-bus fleets outside of China.

Society

Enrique Peñalosa, the former mayor of Bogotá, apart from commissioning TransMilenio to prevent social exclusion, implemented hundreds of kilometres of pedestrian promenades and separated bicycle paths. The construction of public libraries, schools, and nurseries improved the quality of living in poorer neighbourhoods. Efficient public transport system and restrictions on cars improved air quality (Jara-Moreno, 2010).

Human health

The city gained international recognition for outstanding advances in climate action. The city incorporated various strategies to fight air pollution. Bogotá suffered from high concentration of air pollutants, including particulate matter (PM) – dust, dirt, and liquid droplets. Among the threats were short-lived climate pollutant (SLCP) emission from PM emitted from incomplete combustion processes of fossil fuels and biomass. By December 2023, the local government implemented over 20 actions to enhance air quality. These included renewing the public transport fleet with low- and zero-emission technologies and transitioning the manufacturing industry from coal and diesel to natural gas (Pedraza, 2024).

The major objective of the municipality is to ensure a better quality of life within the city. That is a difficult challenge, as Bogotá is among most unequal cities, and the gap between the rich and people living in extreme poverty is deep. The city's strategy was to improve the transportation services and promote walking and cycling in safe environments. In total, 17,000 square miles of street space were repurposed for pedestrian use. Soon after the COVID-19 pandemic began, Bogotá became one of the first cities in the world to create emergency bike lanes (Sustainable Transport Award, 2022).

Asia

Two case studies from Asia were selected to demonstrate the possibilities of developing new towns and districts from scratch, on previously undeveloped land. They both have ambitions to become a showcase of the most advanced ecological solutions. Both of them are interesting examples of the endless struggle of humanity with the unpredictable forces of nature.

NanSha Hengli Island area in China

This development is an example of a new district built from scratch on previously undeveloped low-lying territory, threatened by flooding, water logging, and land subsidence.

The Hengli Island is located in the estuary of the Pearl River, the Pearl Bay, at the tip of Lingshan Island, Mingzhu Bay, Nansha New Area, Guangzhou City, Guangdong Province, Guangzhou's Nansha District, in the Greater Bay Area, the geographical geometric centre of Guangdong Hong Kong Macao Bay area.

Hengli Island is a low-lying territory formed by long-term accretions of sediments from the tributaries of the Pearl River Delta and the tidal erosion and deposition (Dai, 2021). The average elevation of the island is below 1.0 m. The lowest elevation is –3.0 m. A specific phenomenon are daily tidal fluctuations. Due to the influence of the daily tide, the water level of waterways can fluctuate with an average of 1.5 to –3.0 m (Dai, 2021).

Nansha New Area has a total area of 803 km^2 (Zhang et al., 2020). The area belongs to the subtropical monsoon ocean climate with high temperature, high relative humidity, high salt content, strong solar radiation, small temperature difference, and long sunshine. The average yearly temperature is 27°C, ranging from 19°C to 33°C (weatherandclimate.co.uk, 2024a). Annually falls about 1604 mm of rain.

In 2012, Nansha District became the Chinese national new area. It is a small corner of Nansha New Area, located in the middle of Nansha, almost perfectly cantered within the Great Bay Area equidistant from downtown Guangzhou, Hong Kong, Macau, and Shenzhen. NanSha Hengli Island project is born out of an ambition to become a standalone global financial centre – World Financial Island. The new district is built from scratch on previously undeveloped farmland. The construction is envisaged at a very rapid pace. The district is planned to be built in five years. It is stated that the population will increase from 0.72 million in 2018 to 2.25 million in 2030 (Dai, 2021).

Hengli Island is the most important core area of Nansha District. It covers the area of 17.9 km^2 and water area of 0.78 km^2. The waterways around the island create a shoreline of 21.8 km (Dai, 2021).

589.26-ha island provides 130-ha open space with 67% ecological service value increased. The project would create 18 ha of underground space (China-Britain Business Focus, 2020).

Ecology

The most important for Nansha is to improve the blue networks because the rapid urbanization and climate change are constant pressures. The most imminent is the risk of flooding, water backtracking, salty tidal intrusions, and land subsidence. The connectivity of water system and building new corridors is important, so is the preservation of large ecological patches (Nansha Wetland Park). Working with nature is a priority for restoration of natural landscape of mangroves. Design for resilient coastline focuses on improving the functions and creating coastal buffer belts. The pragmatic tradition of local Lingnan culture influenced the project. The principle of traditional local architecture (windy, green, transparent, permeable) was used for planning the district layout. To prevent excessive heat, the shading was used. Energy consumption monitoring was applied to the project. The project is located in the Pearl River estuary, where the Dutch ecological levee with flood control was constructed. The local Lingnan culture adapted to geographical and climatic characteristics of low-lying territory to create water town community.

The project was envisaged as green ecological showroom for urban design. The most important part of landscape of Hengli Island are water canals. They are forming the ecological landscape. They are the lifeline and backbone of the landscape, based on the concept of the sponge city. The sponge technology was applied for rain flood parks and other measures to protect regional wetland ecosystems.

This area becomes a place for research on the construction of high-quality wetland ecosystems in the region. The study is aimed at providing technical support for regional high-quality water ecological construction. The plants were used to create the outer river embankments with free ability to grow. The hard revetment ecosystems were cultivated. In this way, natural ecosystems were re-introduced for ensuring biodiversity. The waterfronts of the riverbanks are planned to become attractive places. The eco-island was designed for eco-tourism. The entire area is monitored for water ecology and early warning of water quality deterioration, algae outbreaks, and foreign biological invasions.

One of the priorities of the new project was water management and improving the quality of domestic water. The project focused on improving the quality of pipe network and providing modern pipeline leakage monitoring system with active prevention and control mechanism. The circular economy, closed-circle zero waste approach to water management, led to the construction of the reclaimed water plant in the sewage plant. The sludge reuse plan was carried out to facilitate the harmless treatment of waste.

Another challenge was soil management. There are numerous problems with the soil in the NanSha Hengli Island, such as poor texture, poor aeration, organic matter, and low potassium content. The ambition was to reuse the soil resources, as regional greening landscape project requires a large amount of planting soil. The soil improvement techniques were used, including improved soil aeration, increasing total potassium and organic matter content, to meet the planting soil standards. The current land use in the region is mostly cultivated areas such as seedlings and fish ponds.

The biodiversity protection is important part of the project. There are more than a hundred species of tropical and subtropical plants. There are more than 20 species of birds, including many endangered species. The region is home to more than a hundred species of fish, including mostly the estuarine fish. The fish in water galleries are mainly pure freshwater fish.

Economy

The district should become Pearl River Bay Area Central Business District and will have its own international commercial bank, futures exchange, and a conference centre. It was chosen to become the International Finance Forum (IFF) permanent conference venue and operation centre. It is a cornerstone tenant. Some of China's most important financial Institutions will have their house branches in the district.

Many financial companies are attracted to Hengli Island (JP Morgan Futures, Postal Savings Bank of China (PSBC) Consumer Finance, and Fosun Health Insurance Company, (China-Britain…, 2020). The multifunctionality, diversity, and multi-use planning were a priority from the very beginning.

Energy resources

The goal is to reduce energy consumption per unit of GDP by 16% and reduce carbon dioxide emissions by 17%. Renewable energy is produced on site using natural conditions – solar energy technology – PV and solar collectors, as well as water resources – e.g., river water cooling. Air source heat pump and water source heat pump are promoted, especially for domestic use.

Energy efficiency methods and new technologies are used to save energy during transmission and consumption. Meeting the energy efficiency standard for new construction was required of all buildings. The energy savings ratio was improved most significantly for public buildings, a bit less for residential construction. The investment necessary for energy-efficient solution was carefully calculated and optimized. The district is equipped with own power supply stations. It is recommended to use light and heat energy storage equipment or increase the utilization rate of solar energy. Solar energy equipment was displayed to promote wider use of PV cells and thermal solar energy.

Regional planning and construction was used to achieve cascade utilization of multiple energy coupling.

Smart City strategy is based on a three-tier structure:

1 Information infrastructure layer – automated data collection and transmission, coordinates the deployment of regional information
2 Information platform layer – collection, integration, analysis, and visualization of data
3 Urban application layer – urban construction, management and operation, smart environment for industry, and better, liveable neighbourhood for residents

The pilot projects include smart buildings, smart street lights, smart driving, and smart streetscapes. It is planned to gradually create a prototype of a regional smart city, with intelligent solutions encompassing communication, energy, environment, urban security, and transportation. The life cycle analysis and supervision is carried out.

Traffic improvements

The public transportation system is planned to include urban rail transit, railcars, motorbikes, and so on. The new rail lines are built to connect the Hengli Island with rapid public transport system.

The Nansha New District is an important transportation hub connecting the both banks of the Pearl River Estuary.

The rule of "10, 30, 60" was used for the integration of multimodal transport.

- 10-minute drive from the place of origin (home or office) should be necessary to reach the highway;
- 30-minute distance between the centre of the district and the main transportation hub; and
- 60 minutes to get to the central city of Guangzhou and the main airports.

To regulate traffic, intelligent parking guidance is used to regulate supply and demand.

The system is successful, the bus sharing rate averages 45%–50% during peak hours.

The district has its own water purification station and a sewage treatment plant to achieve zero-sewage discharge.

Waste management in the new district allows for 50% of the utilization rate of regional domestic garbage. There are numerous domestic garbage collection points, as the service rate is no more than 70 m. From the collection point to the garbage transfer station, the waste is transported via a compressed garbage collection vehicle. After the flammable domestic waste is treated, the non-combustible part would be sent to the landfill. So, there is still much to be done to minimize the waste deposition at the landfill.

The recycling rate of construction waste is planned to be more than 80%. There is a recycling treatment centre for construction waste. In the treatment centre, the waste is crushed, classified, processed, and used to produce non-load-bearing prefabricated components. The recycled waste is used to produce non-load-bearing bricks and square piles for temporary projects.

A resource saving mode was implemented in the entire region, based on incentives and restrain model.

The plan was implemented to optimize the use of silt to reduce the moisture content of the silt and to improve the soil quality for roadbed filling or greening.

Society

The district will have all the necessary public facilities such as hospitals, schools and kindergartens, cultural and community centres, as well as transportation facilities.

The district is attracting high-end living communities. All of the facilities are open to inhabitants and visitors to meet the daily needs: small supermarkets, bookstores, pharmacies, restaurants, and other commercial service facilities. There are numerous places which can become third places, e.g., community centres, both public and commercial.

Human health

In the district, the basic as well as more specialized health services are offered. The basic healthcare helps to deal with common and frequently occurring diseases.

There are also rehabilitation centres. For the elderly there are centres that provide catering, entertainment, fitness, healthcare, and other services. Early warning and prevention systems for urban disaster were introduced with dikes and roads to control the flood.

Special care was taken for ambient air quality and public health promotion. The air pollution is monitored. The dust, domestic gas leakage, and automobile exhausts are controlled. The green belts on roads are constructed to prevent not only air pollution, but also noise pollution. The noise is controlled with sound insulation barriers in sensitive places (acoustically insulated windows).

Masdar City in United Arab Emirates (UAE)

Masdar City is an urban community in Abu Dhabi. The climate is hot desert climate (BWh), according to Köppen-Geiger system. There is no rainfall; the average yearly precipitation rate is 42 mm. The average annual temperature is 27.9°C (Climate Data, 2024b). The city is located at 3 m.a.s.l. (https://whatismyelevation.net/countries/united-arab-emirates).

In 2023, the population in Masdar was 5k people, and the area is 6 km^2.

The story of Masdar City is a valuable lesson that defying the rules of nature and constructing the city in inhospitable desert is extremely costly. Even though the newest advancements in technology and sustainable design were employed, the project is still unfinished. Another lesson is that flexibility and resilience is needed when planning a new development.

Masdar City was designed by Foster and Partners and developed jointly with Abu Dhabi Government. It was built by Masdar, a subsidiary of the Mubadala Development Company – a state-owned holding company of the Abu Dhabi government in the desert next to Abu Dhabi International Airport. The construction of a city in inhospitably desert and not in environment naturally suitable to human life was a challenge, possible only with a large budget. The city was planned to become a sustainable, low-carbon eco-city in the desert. It represents the operationalization of several decades of theoretical research into sustainable development. Masdar was originally intended to be completed by 2016, but, due to the global financial crisis, the date was postponed to 2025.

The important lesson from the history of Masdar is the requirement to be flexible in the pursuit of the sustainable objectives. The primary goals, which were extremely costly, needed to be readjusted due to economic circumstances. Some of the initial concepts, raised construction of the entire city on a platform, and allowing a fleet of computer-driven electric cars to navigate the tunnels under the concrete base, proved to become too expensive. More conventional streets construction and public transport corridors were planned instead (Griffiths & Sovacool, 2020). Despite its disadvantages and changes to original concept, Masdar represents an important advance in the field of sustainable architecture and urban development. The city represents the approach to eco-city as a self-sufficient enclave, reducing reliance on external infrastructures to provide water, food, energy, and other critical resources. Masdar City is employing the most advanced ecological solutions to overcome resources constraints. Abu Dhabi built its economic development on hydrocarbons, like oil, and the UAE still ranks high on a list of countries that had the world's largest ecological footprint per capita. Masdar City demonstrates the effort to shift to renewable sources of energy and other substantial resources and minimize the ecological imprint.

Ecology

Masdar City was the first project of its kind, so there existed no evidence on how to successfully build a sustainable low-carbon city from scratch.

The objectives planned were extremely ambitious. Masdar City was planned to become sustainable, zero-carbon, zero-waste and built on a 9-m raised platform with an electric-powered rapid transport system beneath car-free city. Due to the limitations encountered during the initial implementation, the Masdar City had to reduce its sustainable ambitions. The financial constraints have pushed back the schedule. Nonetheless, the government admitted to this change in objectives and clearly communicated them to the public and fellow stakeholders (https://www.centreforpublicimpact.org/case-study/masdar-city).

Masdar's net energy demands were to be reduced by 70% compared to the average urban city. The city ambition was to rely entirely on solar energy and other renewable energy sources. Today the solar energy production is one of the advantages of the project. The city is operating the largest solar PV plant in the Middle East. The entire community of Masdar City is powered by a 22-ha field of 87,777 solar panels. Additional panels mounted on the roofs of the buildings and projecting over the streets provide an additional source of energy, along with shade for the pedestrians below. The city also envisaged to use geothermal energy, hydrogen energy, and wind energy and develop waste-to-energy projects.

The most scarce resource in the desert is water. The net water demand was envisaged to be reduced 300%. The city water supply was planned to be provided by a desalination facility powered by hydrogen plant. For potable water, the city relies on advanced systems of notification if there are any water leaks throughout the city. Moreover, the installation of high-efficiency fixtures and appliances in the buildings is required. For landscaping, only plants with low water requirements are selected. They would be irrigated with purified wastewater.

The net-waste production was planned to be reduced 400%. Recycling and waste-to-energy production was envisaged.

Very ambitious plan to develop car-free city initially considered that cars would be banned within the city, with available public mass transit and personal rapid transit (PRT) systems. However, the cost of segregating the PRT system from pedestrian traffic hindered the development of the system.

The original plans for the Masdar City have changed multiple times, and the sustainable objectives of the development were reduced due to the financial crisis and drastic shifts in petroleum prices. The business model was changed, and the Masdar City also moved away from a zero-carbon strategy to one focused on carbon neutrality (Griffiths & Sovacool, 2020).

Economy

One of the planned objectives was to "become a regional engine for technology driven growth". Masdar City's ambition was to become a hub for cleantech companies. The city hosts the headquarters of the International Renewable Energy Agency (IRENA). The city was planned to accommodate 1,500 businesses, primarily commercial and manufacturing firms specializing in environment-friendly products. The ambition was to create 10,000 new jobs. More than 50,000 workers were expected to commute to the city daily.

The Masdar Institute of Science and Technology, modelled on MIT, was created inside the city in 2007. It was a research institute focused on clean and renewable energy, water and environment. In 2017, it was merged with the Khalifa University of Science, Technology and Research and the Petroleum Institute to form the new Khalifa University of Science and Technology (Griffiths & Sovacool, 2020).

Masdar City's economic model now positions Masdar as both master builder and real estate developer. As real estate developer, Masdar leases land to third parties. As master builder, Masdar develops buildings and subsequently owns these buildings. The city is built to a demand, depending on clients who wish to occupy the buildings. It is not the traditional approach to build empty buildings and then try to find people to fill them up.

Society

The master planning approach to Masdar City represents an immense and ambitious undertaking. Planned objectives were to accommodate 50,000 residents, 50,000 commuters, and more than 1,500 businesses (Lau, 2012). However, out of the 50,000 people that were planned to inhabit Masdar City, only about 1,300 were actually living in it in 2017 (Omar, 2018).

The Masdar City might be regarded as one of "human-benefits" cities that would secure broad advantages for both the health of the natural environment and the standard of living for all inhabitants. The central principle is that ecologically sustainable communities should simply be attractive cities, with low pollution and greater social solidarity (Lau, 2012).

Human health

The Masdar City was planned according to ecologically conscious urban planning and a "human-benefits" perspective that emphasizes inhabitants' quality of life (Lau, 2012). Evidence-based design, based on research evidence, was used in the innovative urban planning design. Simultaneously, the designers look back at how people had survived in the desert. Traditional Arabian architecture and vernacular methods to cope with hostile desert climate were employed, along with expensive technological solutions and future technology coming on the market.

A remarkable aspect of Masdar City master plan are the three "green fingers" or linear parks that traverse the city and bring cool air. These parks are oriented to direct the prevailing winds into the city. They also provide shade and seating areas and encourage outdoor activities and social interactions. Centrally located public park – Central Park Masdar – is a centre for social life (Property Finder, 2024).

Walking for errands is promoted. Roof projections are casting shadow on the streets. Streets are narrow and orientated to channel the typically north-western breeze. Wind towers are used to provide fresh air (Masdar City was named...).

Australia

The two selected case studies from the Australia are both intentional communities. They demonstrate the possibility of navigating towards a more conscious and sustainable living in a multigenerational communities. They exemplify the model of governance called sociocracy. They may represent the strive to counteract the solitude and anonymity of contemporary life.

Narara Ecovillage

Narara Ecovillage is an "intergenerational residential community" (Narara Ecovillage, 2024a, 2024b). It is an example of group self-build housing in which both social and environmental sustainability are inherent. This ecovillage cooperative enabled project of a similar scale to speculative housing projects (Heffernan & de Wilde, 2020). There are numerous advantages of this project: energy efficiency, affordability, quality, innovation, sustainable communities, and meeting the needs of occupants. The process also allowed for the ability for self-builders to share information and raise awareness of zero carbon (Heffernan & de Wilde, 2020).

Narara Ecovillage is located in Narara, a suburb of Gosford, a city on the Central Coast of NSW, Australia. The city of Gosford is located just over an hour drive of Sydney. The climate type is of temperate coastal regions around Sydney and the Central coast (NSW Government, 2024). The average monthly temperature is 18°C–28°C, and the average yearly temperature is 23°C. The precipitation is averaging 123 mm per month and 1481 mm per year (Weathermondo.com, 2024).

This Ecovillage is an intergenerational residential community, constructed on the site of a former horticultural research station. The site measures 63 ha – 12 ha are developed as residential, 12 ha are used as common gardens and agriculture, and 39 ha are covered with native forest and bushland (nararaecovillage.com).

The project started in 1999, and the first houses were completed and first inhabitants moved in 2019. There are currently 180 people living on site (Mottl, 2021). The vision of Narara Ecovillage is for an environmentally, socially, and economically sustainable world. The development is phased in three stages. The village is planned to house more than 300 people in more than 150 homes (Collaborative Housing, 2024).

Ecology

There are numerous ecological advantages that come with the self-build projects. The community envisaged high levels of sustainability. The ecovillage has developed and applied various principles during the planning stages, which are simple yet effective steps towards making a difference to the environment. All the buildings in the village must meet the standards of sustainable construction, which is verified by trained assessors and Building Review Panel of the ecovillage. Among the aspects which are considered are local materials, embedded carbon, and water efficiency. All types of construction are available – hempcrete, straw-bale, and rammed earth, as well as concrete and lightweight frame construction. The individual owners are encouraged to build smaller, inexpensive houses with low maintenance cost and use innovative solutions. The use of local, sustainable solution is promoted. One of the advantages of self-build community is willingness to explore lower impact materials and methods. Moreover, lower cost frees up funding for additional measures. The community has the capacity to "spend to save", and self-builders have a vested interest in quality (Heffernan & de Wilde, 2020).

Some other principles included: situating houses in the natural setting of the native forest and replacing traditional fencing with shrubs and trees. The community is trying to maintain and regenerate the native bushland on the property by replanting local species in place of removed invasive ones. The property is surrounded by natural Strickland Forest. One of the disadvantages of ecovillages is that the sites are difficult to obtain (Croft et al., 2022).

In addition to these simple principles, the ecovillage has cutting edge technology for both power and water supply to its residents. The community has its own smart grid and manages its own potable water. The ecovillage has been granted a community-owned water license to supply drinking water from the dam. The water is supplied by a creek which flows through the ecovillage – one of the tributaries of Narara Creek and supplemented by catchments in the surrounding Strickland State Forest. The damn has a capacity of 45 million litres. The license also allows for the treatment of waste water. The treated wastewater is used for toilet flushing, gardening, and other agricultural uses. The community is 100% water self-sufficient (Ghosh, 2024; Narara Ecovillage, 2024; Stibel, 2022).

The food production on the site includes: mixed fruit and nut orchards, community gardens, seedling propagation, chooks, and beehives. There are a few experimental beds in the greenhouse. Around 12 ha are used for agriculture and community gardens (Earth haven, 2024).

Economy

The ecovillage is pushing the boundaries of conventional building, water and power models. Narara Ecovillage aims for energy self-sufficiency through mandating a net-positive energy target for each house and the development of a smart grid. Moreover, it plans to be carbon neutral or even carbon positive. The village has its own subsidiary NEV Power, a community smart grid which provides power to Narara Ecovillage and Narara Ecovillage Cooperative. It is an example of embedded two-way energy network. It is one of the advantages of living in a community group – the opportunity for owners to manage their local generation and pool output and storage resources. The NEV Power mission is to maximize the supply and utilization of renewable energy. It is expected to have relative low consumption and high surplus of energy to export. All the owners of houses in ecovillage must install sufficient PV panels to meet their own needs and produce a surplus. The minimal surplus is calculated to cover the energy embedded in the construction of their houses. However, most owners installed much larger systems. The NEV Power is connected to the national grid to guarantee the reliable voltage across the network and to export the surplus of energy production. Moreover, a battery storage system is ready to provide the community with the energy in times when solar energy production is limited. The individual customers are also advised to invest in a battery for their needs and introduce their own control systems. The NEV Power has a smart grid Wi-Fi network to supervise the energy production and consumption in real time. Smart grid technology is used to monitor energy loads, determine peak load periods, and forecast weather events that can affect energy production. The community pooling of individual energy production and storage allows optimizing the generation, storage, and load distribution (Australian Government, Undated). For the hot water, individual heat pumps are proposed, as well as a range of ventilation with heat recovery to be installed in each home (Narara Ecovillage, 2024; Stibel, 2022). One of the advantages of self-build community, apart for shared energy sources, is better understanding of the home energy system, ability to specify higher standards of energy efficiency, and sharing information and awareness (Craft et al., 2022).

For the transportation, the ecovillage is prioritizing pedestrian and cycling access around the village with vehicle parking kept to the outskirts of the property.

The advantages of self-build community are affordability and lower running costs. They can benefit from economies of scale in the procurement of goods and services and the purchase of land. By undertaking elements of the work by the community members themselves, they could make additional savings. The economies of scale include shared energy systems, car sharing, food growing, and bartering of skills (Craft et al., 2022).

Society

This project is an intergenerational residential community. It was designed to be a showcase of friendly and stylish ecovillage. The project is focused on the well-being of the inhabitants and our planet. The members hold individual ownership of their homes as well as a share in the cooperative community owned premises. Members contribute 1 hour of work to cooperative maintenance and development. The village was designed for ageing in place, and all age groups are welcomed. The adaptable design is thus promoted. Majority of members are over 50 years old, but all age groups are present, including children (Families on the coast, 2021; Narara Ecovillage, 2024a, 2024b).

The governance model is based on sociocracy (dynamic governance), where everyone has an influence on decision-making. This model is adapted in many ecovillages around the world. One of the strategic goals is to live with kindness, aware of interdependence with others and the natural world (Narara Ecovillage, 2024a, 2024b). The residents and visitors are served at an on-site coffee cart. The community is organizing meetings, presentations, workshops, and other events. The smart-grid NEV Power Ltd is offering local job opportunities in new renewable technology.

One of the major advantages of self-build community is the social capital which is created. The skills and confidence gained through the self-build have potential for the empowerment of the individual, the group, and the local community. The act of building, designing, planning, or procuring as a group creates bonds within a community. Self-builders are typically interested in sustainable lifestyle choices. The self-build communities are more resilient and less transient. They are supporting each other, and they support local businesses. The self-build community meets the needs of occupants, gives ability to procure a tailored design and increased satisfaction (Craft et al., 2022). One of the most outstanding achievements of Narara Ecovillage was its ability to harness the collective power of the community by maintaining a clear and shared vision of potential to overcome significant financial and regulatory challenges in making NEV's vision a reality. Its members are not only inhabitants of the ecovillage but also the investors, developers, decision-makers, and utility providers (Craft et al., 2022).

There are also some disadvantages of group self-build communities. Group issues, reaching consensus, and finding people to collaborate with different lifestyles are among the most important. The project requires buy-in from the stakeholders. On the side of the community member it requires commitment and is time consuming. There are many concerns raised by prospective members: legal complications of shared services and responsibilities (renewables) and the attitude that eco-technology is complicated and unproven (Heffernan & de Wilde, 2020).

Human health

Well-being and good health are the foundation of this project. The healthy environment was the objective of the project. The ecological sustainability is paired with social sustainability. It is a community that provides many opportunities to share activities with neighbours and create social bonds. There are individual houses and cohousing homes in the village.

Jagera EcoCommunity (MacLean EcoVillage)

Jagera EcoCommunity is located in Maclean, in Yaegl Country, Clarence Valley, New South Wales, Australia. The Köppen climate type is Cfa: humid subtropical climate. The population of Maclean is 2,778 (mindat.org, 2024). The summers are warm, humid, and wet; the winters are short and cool. The temperature over the year typically varies from 8°C to 28°C and is rarely below 3°C or above 31°C. (Weather Spark, 2024) The average elevation is 9 m.a.s.l. During heavy rains the town is under threat of flooding by the Clarence River. This location allows for an easy access to the historic Maclean township. The villagers can walk or cycle for everyday shopping and services, e.g., groceries, a cup of coffee, a swim at the pool, or to the local market. The ecovillage is located just 20 minutes' drive to Yamba, magnificent beaches, and the Yuraygir National Park and about an hour drive to the closest airport (Jagera EcoCommunity, 2019, 2023).

Jagera EcoCommunity is a small intergenerational intentional community nestled in forest on a sunny hillside overlooking Birriimba (Clarence River) and the Maclean township. The core site is 2 ha (5 acre) and used to be a paddock in the 1970s. Today, 2.5 acres are dedicated to rainforest restoration, community orchards and gardens, as well as shared facilities. 2.5 acres are dedicated to 11 community title lots. Even though it is so small, this eco neighbourhood is an accredited member of the Global Ecovillage Network (Bragg, 2019).

Jagera ecovillage is using permaculture principles and design. The community has a vision to live with a small ecological and resource use footprint. The aim is to be a model of a resilient and sustainable way of living for the future. The educational aspects are important as the ecovillage is organizing workshops, seminars, and tours to promote regenerative lifestyle and contribute to wider community (OzGreen, 2022).

Ecology

The property is 5 acres of north facing slope nestled in a hidden valley above the town, surrounded by regenerating rainforest. The formed paddock in the 1970s became overgrown with woody weeds. Today it is a zoned residential. The property bounds the Foresthaven wildlife sanctuary to the east and suburban backyards to the north and west (Jagera EcoCommunity, 2019, 2023). Jagera EcoCommunity is designed to be interdependent with Foresthaven eco-retreat and learning centre, adjoining the Jagera property. The Foresthaven has 15 acres of regenerating forest and open parkland, with walking trails leading along the creek. The intermittent rainforested creek runs along bottom of property to the north. It has been regenerated by the founding members of the EcoCommunity for the last 25 years. The results of the rainforest regeneration created a forest rich with native birds and animals. There are some endangered species present at the site. The Foresthaven and Jagera are separate entities, but there are possibilities for Jagera members to facilitate and attend workshops and events at the centre. In Foresthaven, there are two "recycled" weatherboard cottages that are currently available for rent. Moreover, Jagera members are co-creating Foresthaven and regenerating the rainforest.

The members of the EcoCommunity share a common vision for the future, as they declared in their draft vision and values (Jagera EcoCommunity, 2019, 2023). They want to care actively for the Earth, maximizing their positive ecological impact, healing and regenerating natural vegetation, providing habitat for native animals, and retaining water in the landscape (Jagera EcoCommunity, 2019, 2023).

The ecovillage is designed and maintained according to the principles of permaculture – "caring for Earth", "caring for each other", and "fair share". The permaculture system forms the basis for food production (community gardens, chickens, orchards, and bushfoods); movement; water and energy systems (Jagera EcoCommunity 2019, 2023, Bragg, 2019).

Prior to the commencement of designing rigorous flora and fauna, assessments were conducted. Their results were approved by the local authorities (Clarence Valley Council) and used as a basis for the development plans and Neighbourhood Management Statement. The emphasis was on minimizing the impact during construction and the protection of particular areas of ecological and cultural value. The rainforest creek and high sandstone areas are protected, and their ownership is shared. The wildlife corridors of microbats, birds, and terrestrial vertebrates are protected during the construction. The Jagera is treated as a wildlife sanctuary, where the local native animals are safe and free to roam. Therefore, there is a "no dog and no cat" policy as part of the Neighbourhood Management Statement. Part of site management and development is dedicated to biodiversity enhancement strategy and mechanical removal of invasive species. The rainforest regeneration is treated as one of the community activities.

Economy

Jagera Ecovillage is a self-build community. The self-build stipulates smaller houses. Moreover, the Neighbourhood Management Statement encourages small dwelling size, sustainable and cost-efficient construction materials and methods. The Jagera members are inspired to build tiny houses and expanded dwellings if necessary, as well as to share ownership of residential lots. The aim is to reduce the impact of large building structures. Therefore, cottage-style and innovative forms are promoted. Each house is individually designed and built by its owners, but they all have to be in a range of compatible forms using sustainable building materials. The focus is on passive solar design, energy- and water-efficient. The goal is to reduce the ecological footprint of each of the inhabitants of the EcoCommunity. The members of the intentional community are striving for a sustainable and regenerative lifestyle (Sustainable futures Australia Our Eco-Neighbourhood, 2024).

Each member can design and build his own sustainable house. The intention of the EcoCommunity is to be as "off-grid" as possible. These principles are to be applied to solar power, water supply, and wastewater treatment. The community may rely on the safety features of direct access to town utility supplies if needed. Each building will have solar panels and batteries. They all would be connected to the main microgrid operating between the households, sharing power in the community prior to exporting excess power back to the grid. Thus, the households will be able to receive energy when batteries are low, and there are even possibilities to earn some extra money for the community.

All buildings will be designed to take advantage of the north facing aspect of all lots and summer breezes from the river. The passive heating and cooling will be used. For heating in winter and cooking, the usage of timber cleared from the site is envisaged. The use of biogas and micro hydroenergy in times of high creek flow is also researched (Jagera EcoCommunity, 2019, 2023).

Use of private vehicles will be discouraged. The ecovillage is located within walking distance from the town centre with all the necessary shops and services (cheap goods and services, reduced vehicle costs). The community is also discussing shared, solar-powered vehicles.

For the water supply, each household will have its own rainwater tanks on the roof. The roof surface and the tanks volume should be calculated for individual household needs. In case of need, this could be supplemented by the connection to the municipal water system. The rainwater from the internal ring road would be treated and reused for community gardens, orchards, and landscaping. Each household will install grey water systems to reuse household water (Jagera Eco-Community, 2019, 2023).

The community would grow its own food. Gardens and food systems will be designed in detail using permaculture and collaborative principles to suit the evolving needs of villagers. There are community gardens and orchards. Each household would have its own private garden and orchard. If more land would be needed to feed the community, the suitable agricultural area of the Foresthaven other neighbouring properties could be used. The Jagera EcoCommunity is supporting the local farmers markets, as both vendors and customers. No food wasting would be supported. The Neighbourhood Management Statement would help to share meals, preserve food, and communal cooking. The plans are to build a communal kitchen, pantry, cellar, and pizza oven (Jagera EcoCommunity, 2019, 2023).

The EcoCommunity has sustainable building guidelines for the materials used for construction and on site. The guidelines encourage the use of sustainable building materials, including recycled materials, hempcrete, and rammed earth. Today, the neighbourhood is an interesting mix of sustainable residential housing.

The sourcing of materials as locally as possible is also promoted. The use of recycled materials from all the demolition, earthworks, excavation, and construction is researched (road building materials, pipes, and bridge construction materials). All the soil from the excavations would be used onsite. The community intends to minimize the loss of any organic matter from the site. All the wood from the site would be used for construction, mulched or stored as firewood. Sharing of tools and other mechanical equipment is also envisaged.

Waste production should be minimized. The community is connected to municipal sewage system. However, each household will have composting toilets to reduce the sewage load. Compost will be used to fertilize the fruit trees and food forest. The household food wastes would be composted for fertilizing individual gardens, community gardens, and/or biogas generation.

When it comes to financing, the members of Jagera EcoCommunity are able to borrow money from a financial institution. The loan could be based on the individual lot ownership title as financial security. If they decide to leave the community at some point in the future, their investment would keep pace with the surrounding mainstream property market (Jagera EcoCommunity, 2019, 2023).

The growth of Foresthaven as an eco-retreat and workshop venue would create opportunity for the Jagera community members to rent rooms to workshop participants. There are plans for promotion of educational tourism. The mission of the community is also initiating local economic enterprises, food systems, and creative arts (Jagera EcoCommunity, 2019, 2023).

Society

Jagera EcoCommunity is an intentional self-build community and should attract community-minded people. The objective is to support one another in creative endeavours, well-being, and prosperity. Similarly to Narara EcoVillage, the Jagera members hold individual ownership of their homes, as well as a share in the cooperative community owned premises. They all sign and observe the Neighbourhood Management Statement. The governance model is based on collaborative decision-making, facilitated by co-founders, who are experienced in participatory and collaborative planning. It is based on sociocracy, where everyone has an influence on decision-making in face-to-face meetings or online (Jagera EcoCommunity, 2019, 2023). The community is promoting diversity of ages, cultures, financial wealth, interests, and skills (Jagera EcoCommunity, 2019, 2023).

The important aspect of Jagera EcoCommunity is the collaboration with the descendants of the first nations – the Aboriginal community – Yaegl Traditional Owners. For the Cultural Heritage Assessment of the land, the Yaegl consultants were engaged. They would supervise the construction to ensure no damage is done to cultural assets. Yaegl people are employed in construction of community infrastructure and maintenance of the land. The goal is to respect and learn from the Yaegl culture and local Aboriginal community.

The intentional community is encouraging at the same time both privacy and autonomy, as well as sharing of resources, community meals, and gatherings.

One of the missions of the EcoCommunity is engaging with, respecting, and caring about people in our street, neighbourhood, town, region, and rest of the world (Jagera EcoCommunity, 2019, 2023).

Human health

The members of the Jagera EcoCommunity are practicing deep adaptation and regenerative lifestyles. They focus on enhancing well-being, including psychological and social resilience. They practice deep connection with nature and intergenerational collaboration with other community members. The community is encouraging ecotherapy and "forest bathing". Living in an intergenerational community might be a shield against solitude, and thus many health issues that are related to isolation and rejection. The Jagera Community aims to promote caring for each other, joy, fun, and friendship (Jagera EcoCommunity, 2019, 2023).

The members of the community are encouraged to produce their own food.

Africa

Figures suggest that by 2035, more than half of Africans will live in urban environments (Botham, 2020). The two case studies are examples of interventions in different scale. Cape Town is on its way to become a sustainable city, offering a better quality of living to all inhabitants. Castle Landmark is located within the New Administrative Capital city. It is an example of a new neighbourhood developed in a new town – New Administrative Capital of Egypt.

Cape Town (Kaapstad)

Cape Town is a national legislative capital of South Africa. It lies at the coastline, close to the Cape of Good Hope. The city is located between the Table Mountain (1086 m.a.s.l.) and neighbouring peaks and the flat areas below the slopes. The climate is warm Mediterranean climate (Köppen: Csb). The average high temperature is 21°C in January and February, and the average low is 13°C in July. On the average, rain falls on 69 days of the year; the annual rainfall is 660 mm. Cape Town enjoys mild, moderately wet winters and dry, warm summers. Cape Town has an approximate area of 2.5 km^2 and more than 4.4 million inhabitants (Axelson, 2024; Alexander, 2023).

Ecology

The city of Cape Town was hailed the greenest city in Africa (Botham, 2020). There's been plenty of investment in beautifying and maintaining the city's precious green spaces.

Cape Town has been rated as the most sustainable city in Africa according to the Corporate Knights Sustainability Cities Index report released on June 15, 2022. The most important indicators that are measured are: air quality, emissions, solid waste generation, climate change resilience, and sustainable policies. The cities' commitment to renewable energy, reduced GHG, and clean transportation is also important (www.capetowngreenmap.co.za, 2022; Smith, 2022). Cape Town's municipality is committed to becoming carbon neutral by 2050 (Smith, 2022). To achieve this goal, the city is incentivizing green industries, ensuring procurement is green, and generating or purchasing renewable energy.

Economy

The city positions itself as the green economy hub for South Africa. The City's Spatial Development Framework incorporates a major focus on building a sustainable, climate resilient; and resource-efficient city. Some of the changes instigated in the city are driven by academic institutions and non-governmental organizations (NGOs). The important factors were also water and energy shortages. The city has implemented GHG emission reduction policies and strives for energy and water efficiency (Holgate, 2007).

Numerous renewable energy companies have their headquarters in Cape Town. Economic clusters include solar PV manufacturing. The energy prospects for the Cape Town are envisaging energy-efficient small-scale embedded generation and energy storage. The generators of less than 1 MW can be connected to the electricity distribution network to contribute surplus power (Invest Cape Town, 2024). One of the largest sectors for the green building industry is the retrofitting of the country's existing building with high-efficiency lighting, HVAC, and smart metering.

The city authorities have invested in wind energy. The city boasts the first commercial wind farm in South Africa, which has been in operation since 2008. The city hopes to get 10% of its energy from renewable sources (Botham, 2020).

A constrained water supply is one of the major problems for the city. Therefore, numerous initiatives are tested for industrial water reuse, recycling, and resource recovery. Groundwater resources, desalinated seawater, and treated wastewater are the only water sources on a municipal scale.

The city invests in cleaner public transport and a network of cycle lanes (MyCiTi transportation system).

The city of Cape Town is recognizing the importance of prioritizing a green economy, where decent green jobs are created and resilience is improved. The city has implemented environmental fiscal reform initiatives and various other incentives, e.g., launching of Atlantis Greentech Special Economic Zone in 2018. The city and the entire province are offering abundance of natural resources, modern infrastructure, skilled workforce, and relatively low operational costs, as well as prime locations.

Society

Cape Town is highly unequal according to international standards; however, the Gini coefficient of inequality is 0.58 – lower than South Africa's average of 0.7.

In 2009, the city mayor signed a Responsible Tourism Charter, in which the city pledged to follow responsible tourism practices. Responsible tourism creates economic benefits for local people and helps communities develop. The policy of responsible tourism, which the city adheres to, involves local people in decisions that affect their lives. Responsible tourism is culturally sensitive and instills a sense of pride in local people.

Human health

The inhabitants have easy access to healthy food. The locally grown food which is healthier and cheaper than imported is available at the local farmer's markets. Moreover, there are plenty of places to grow food.

Castle Landmark, New Capital Compound in Egypt

Cairo has the hot desert climate (BWh) according to Köppen-Geiger classification, with mean annual temperatures ranging between 9°C and 34°C. Rainfall is sparse. The metropolitan area of Greater Cairo is the largest in Africa. A population is nearly 20 million.

Castle Landmark is a new neighbourhood of the New Administrative Capital of Egypt, a new satellite city of Cairo City. The new city – New Administrative Capital – is intended to serve as the new capital of Egypt. It is to be built further east of the existing satellite city of New Cairo, 45 km east of Cairo.

The New Administrative Capital is designed by Skidmore Owings and Merrill. It is built on a largely undeveloped area between Cairo and Suez. It is planned to have a population of 44 million by 2050, and 21 residential districts (Lang, 2017). One of these districts will be Castle Landmark (New Capital Compounds, Castle Landmark New Capital, 2024).

Ecology

One of the biggest problems of Cairo City is overpopulation. The New Administrative Capital is built to relieve this problem. The advertisements for new apartments in the Castle Landmark are boasting about serenity and relaxation. The designed greenery and artificial lakes are making 81% of the entire landscape. Only 19% of land is dedicated to development (Egypt Real Estate Hub…).

The parking space is located underground to protect the vehicles from excessive heat.

Economy

The new capital is considered one of the projects for economic development. It is supposed to relieve the congestion in Cairo. The New Administrative Capital is built as a smart city. The AI will be used to monitor water use and waste management.

Castle Landmark is developed as a multifunctional neighbourhood. The project includes the everyday life facilities and amenities, e.g., post office, markets, shops, nursery, leisure places, sports facilities, swimming pools, gyms, dining destinations, and much more. The project area is 43 acres (0.174 km^2).

The neighbourhood consists of 44 buildings, 1,800 housing units of various sizes: 80–285 m^2. There are no villas offered in the neighbourhood (Egypt Real Estate Hub…).

The solar energy is satisfying 70% of needs when it comes to lighting buildings and greenery in the compound.

The New Administrative Capital is connected with Cairo through the Cairo Light Rail Transit (LRT) and a monorail line. The Castle Landmark is located 32 km from Cairo Airport, 20 minutes away from the airport of the New Administrative Capital.

There are various possibilities for payment plans for the apartments in Castle Landmark.

Society

The criticism is formulated as the new capital is considered a city for the rich. The prizes of the apartments are out of reach of the middle and lower income groups. The assumption that the city for the rich is financed with public money inspires even more criticism.

The project was designed with people with special needs in mind. The inclusive design of all the entrances and inner roads ensures enjoyment and practicality to all.

Human health

The Castle Landmark was designed to offer the best quality of living and well-being to its inhabitants and visitors. All the apartments have a view of a garden, greenery landscape, or lakes. The homes are designed to enjoy plenty of sunlight. The Castle

Landmark is located close to the cathedral church. Medical services are available to Castle Landmark inhabitants.

The privacy, security, and protection from stressful hazards (noise, pollution, overcrowding, etc.) are all important factors.

References

AFD (Agence Française de Développement). Undated. Supporting sustainable urban mobility in Curitiba, available online: https://www.afd.fr/en/carte-des-projets/supporting-sustainable-urban-mobility-curitiba retrieved on 30.04.2024.

Alexander Ch. 1977. *A Pattern Language. Towns – Buildings – Construction.* Oxford University Press, New York.

Alexander, Ch. 2002. *The Nature of Order: The Process of Creating Life*. The Centre for Environmental Structure, Berkeley, CA.

Alexander, M. 2023, South Africa's weather and climate, South Africa Gateway, available online: https://southafrica-info.com/land/south-africa-weather-climate/, retrieved on: 27. 12.2024.

Australian Government. Undated. Narara Ecovillage smart grid, available online: https://arena.gov.au/projects/narara-ecovillage-smart-grid/ retrieved on 29.06.2024.

Axelson , E. 2024 Cape Town national legislative capital, South Africa, Encyclopedia Britannica, available online: https://www.britannica.com/place/Cape-Town/The-people, retrieved on 27.12.2024.

Barrero K. 2023. Bogotá contará con dos nuevos ecobarrios en San Cristóbal y Usme Noticia de Septiembre 23, 2023, available online: https://bogota.gov.co/mi-ciudad/habitat/bogota-contara-con-dos-nuevos-ecobarrios-en-san-cristobal-y-usme, retrieved on 23.05.2024.

Bernal C., Durosaiye I., Hadjri K., Corredor S. Duran E., Prieto A. 2022. Neglected landscapes and green infrastructure: The case of the Limas Creek in Bogotá, Colombia. *Geoforum*, 136, 194–210. https://doi.org/10.1016/j.geoforum.2022.09.010.

Botham D. 2020. The world's greenest cities: Cape Town, South Africa, published January 14, 2020, available online: https://www.litterbins.co.uk/blogs/news/the-worlds-greenest-cities-cape-town-south-africa retrieved on 13.07.2024.

Bragg E. 2019. Jagera EcoCommunity, global ecovillages network, available online: https://ecovillage.org/ecovillage/jagera-ecocommunity/ retrieved on 12.07.2024.

Brodzinsky S. 2014. A tale of two cities: Bogotá, *Americas Quaterly*, January 25, 2014, available online: https://www.americasquarterly.org/fulltextarticle/a-tale-of-two-cities-bogota/ retrieved on 01.05.2024.

Cabral P. 2022. Curitiba, Brazil: A sustainable city, *CGTN*, America, available online: https://america.cgtn.com/2022/10/10/curitiba-brazil-a-sustainable-city retrieved on 11.04.2024.

China-Britain Business Focus. 2020. *Nansha's Hengli: World Financial Island*, published on 28.01.2024, available online, https://focus.cbbc.org/nansha/amp/amp/, retrieved on 01.06.2024.

City of Ithaca. 2024a. City of Ithaca Energy Action Plan (EAP) 2012–2016 (PDF), retrieved on 27.04.2024.

City of Ithaca. 2024b. Local action plan: To reduce greenhouse gas emissions for City of Ithaca Government Operations (PDF), retrieved on 27.04.2024.

City of Ithaca. 2024c. Resolution, available online: https://www.cityofithaca.org/422/City-Commitments-Resolutions retrieved on 27.04.2024.

Clichy-Batignolles. Undated. (Paris 17th) A new urban quality for northwest Paris, available online: https://www.parisetmetropole-amenagement.fr/en/clichy-batignolles-paris-17th#scrollNav-4 retrieved on 17.04.2024.

Climate Data. 2024a. Athens Climate (United States of America), available online: https://en.climate-data.org/north-america/united-states-of-america/georgia/athens-1642/ retrieved on 01.07.2024.

Climate Data. 2024b. Climate: Abu Dhabi, available online: https://en.climate-data.org/asia/united-arab-emirates/abu-dhabi-2163/ retrieved on 01.07.2024.

Climate Data. 2024c. Curitiba Climate (Brazil), available online: https://en.climate-data.org/south-america/brazil/parana/curitiba-2010/ retrieved on 01.07.2024.

Climate Data. 2024d. Mazatlán Climate (Mexico), available online: https://en.climate-data.org/north-america/mexico/sinaloa/mazatlan-871696/ retrieved on 01.07.2024.

Collaborative Housing. Undated. Narara Ecoviallage, available online: https://www.collaborativehousing.org.au/stories-narara-eco-village retrieved on 29.06.2024.

Costa E. Silva J.A., Steffen R.E. 2019. Urban environment and psychiatric disorders: A review of the neuroscience and biology. *Metabolism* 100S, 153940. https://doi.org/10.1016/j.metabol.2019.07.004.

Croft W., Ding L., Prasad D. 2022. Understanding decision-making in regenerative precinct developments. *Journal of Cleaner Production*, 338, 130672. ISSN 0959-6526. https://doi.org/10.1016/j.jclepro.2022.130672

FCI (Flora Culture International). undated. Curitiba, Brazil: Curitiba's Urban Agriculture, available online: https://aiph.org/green-city-case-studies/curitiba-brazil-urban-agriculture/ retrieved on 30.04.2024.

Dai, W. 2021. Spatial Planning and Design for Resilience: The Case of Pearl River Delta. [Dissertation (TU Delft), Delft University of Technology]. https://doi.org/10.7480/abe.2021.04.

Data USA. Ithaca, NY, available online: https://datausa.io/profile/geo/ithaca-ny/ retrieved on 01.07.2024.

Dobbs C., Hernandez-Moreno A., Reyes-Paecke S., Miranda M.D. 2018. Exploring temporal dynamics of urban ecosystem services in Latin America: The case of Bogota (Colombia) and Santiago (Chile). *Ecological Indicators*, 85, 1068–1080. ISSN 1470-160X, https://doi.org/10.1016/j.ecolind.2017.11.062.

Earth haven. 2024. Narara Ecovillage Masterplan, available online: https://earth-haven.com/narara-ecovillage-masterplan/ retrieved on 11.07.2024.

Egypt Real Estate Hub. Castle Landmark – New Capital, available online: https://www.egyptrealestatehub.co.uk/castle-landmark-new-capital retrieved on 17.07.2024.

El sol de Mazatlan. 2021. Termina la primera etapa del Parque Aeroespacial en Mazatlán, available online: https://cobertura360.mx/2021/12/11/sonora/termina-en-mazatlan-la-primera-etapa-del-parque-aeroespacial/ retrieved on 29.04.2024.

Families on the coast. 2021. Narara Ecovillage: Where families thrive in community, sponsored editorial published on June 1, 2021, available online: https://onthecoastpublications.com.au/onthecoastfamilies/narara-ecovillage-where-families-thrive-in-community/ retrieved on 11.07.2024.

Francia reafirma su compromiso de cooperación con "ecobarrios" de Colombia. 2023. SWI swissinfo.ch, 24.01.2023, available online: https://www.swissinfo.ch/spa/francia-reafirma-su-compromiso-de-cooperaci%C3%B3n-con-ecobarrios-de-colombia/48230448, retrieved on 2.5.2024.

Ghosh S. 2024. Water management practices in master-planned developments and eco-neighborhoods. In: Abraham M.A. (Ed.), *Encyclopedia of Sustainable Technologies*

(2nd ed.), Elsevier, Amsterdam, Oxford, Cambridge. pp. 803–819. ISBN 9780443222870, https://doi.org/10.1016/B978-0-323-90386-8.00074-7.

Global BRT Data. 2024. Available online: https://brtdata.org/#/location retrieved on 30.04.2024.

Gnatek T. 2003. Brazil-Curitiba's urban experiment, Frontline World, available online: https://www.pbs.org/ frontlineworld/fellows/brazil1203/ retrieved on 30.04.2024.

Gobierno de Mexico. 2024. Mazatlán, available online: https://www.economia.gob.mx/datamexico/en/profile/geo/mazatlan retrieved on 01.07.2024.

Grave R. 2021. Parque Aeroespacial y MLC fortalecerán al sector educativo. Se acabó la fuga de cerebros en el Puerto, Punto MX, 1.12.2021, available online: https://punto.mx/2021/12/01/parque-aeroespacial-y-mlc-fortaleceran-al-sector-educativo/ retrieved on 29.04.2024.

Griffiths S., Sovacol B.K. 2020. Rethinking the future low-carbon city: Carbon neutrality, green design, and sustainability tensions in the making of Masdar City. *Energy Research & Social Science*, 62, 101368. https://doi.org/10.1016/j.erss.2019.101368.

GPA (The Georgia Planning Association). 2024. Official Chapter of the American Planning Association (APA), *2023 GPA Fall Chapter Awards*, available online https://georgiaplanning.org/wp-content/uploads/GPA-Awards-2023.pdf retrieved on 27.04.2024.

Guislain M. 2016. Le macrolot, paysage urbain du XXI siécle? *AMC Le Moniteur Architecture*, 2016, 855.

Guislain M. 2019. Christian de Portzamparc, en noir & blanc et en lumière. *Le Moniteur*, 30 decembre 2019, available online: https://www.lemoniteur.fr/article/christian-de-portzamparc-en-noir-blanc-et-en-lumiere.2069424 retrieved on 15.09.2023.

Heffernan E., de Wilde P. 2020. Group self-build housing: A bottom-up approach to environmentally and socially sustainable housing. *Journal of Cleaner Production*, 243, 118657. ISSN 0959-6526, https://doi.org/10.1016/j.jclepro.2019.118657.

Hidalgo, D. 2014. Urbanism Hall of Fame: Jaime Lerner – The Architect of Curitiba available online: https://thecityfix.com/blog/urbanism- hall-fame-jaime-lerner-architect-curitiba-dario-hidalgo/ retrieved on 24.02.2015.

Holgate, C. 2007. Factors and actors in climate change mitigation: A tale of two South African cities. *Local Environment*, 12(5), 471–484. https://doi.org/10.1080/13549830701656994.

Industrial Park. undated. MZT Aerospace, available online: https://aerospacepark.mx/ retrieved on 29.04.2024.

Invest Cape Town. 2024. Cape Town makes serious strides towards a green economy, available online: https://www.investcapetown.com/why-cape-town/green-economy/ retrieved on 13.07.2024.

ITDP (Institute for Transportation and Development Policy). 2009. Lessons of urban transit from Bogota, available online: https://www.itdp.org/lessons-of-urban-transit-from-bogota/ retrieved on 27.05.2024.

Jagera EcoCommunity. 2019. Application for membership of global ecovillage network Australia, available online: https://static1.squarespace.com/static/5c367f6ec3c16a98d012bb9c/t/5c528151c74c50734e5e5137/1548910933981/Jagera+Ecovillage+Accreditation+15+Jan+2019.pdf retrieved on 12.07.2024.

Jagera EcoCommunity. 2023. Available online: https://www.ecocommunity.org.au/ retrieved on 12.07.2024.

Jara-Moreno A. 2010. TransMilenio celebrates 10 years of operation, published on 03.11.2010, available online: https://www.itdp.org/transmilenio-celebrates-10-years-of-operation/ retrieved on 27.05.2024.

Jenkins P., Smith H. 2024. *Order and Disorder in Urban Space and Form.* Routledge, Abingdon, Oxon, New York.

JHP. 2021. North Downtown Athens master planning study redevelopment masterplan for Bethel village/College & Hoyt, Athens, March 2, 2021, available online: https://ndathens.com/wp-content/uploads/2022/03/2019069_210301_Feb-final-report_Small.pdf retrieved on 27.04.2024.

JHP Architecture/Urban Design. 2023. North Downtown Athens master plan wins GPA award, available online: https://jhparch.com/north-downtown-athens-master-plan-wins-gpa-award retrieved on 27.04.2024.

Kottek M., Grieser J., Beck C., Rudolf B., Rubel F. 2006. World map of the Köppen-Geiger climate classification updated. *Meteorologische Zeitschrift*, 15(3), 259–263. https://doi.org/10.1127/0941-2948/2006/0130.

Lang J. 2017. *Urban Design.* Routledge, Abingdon, Oxon, New York.

La préfecture et les services de l'Etat en région l'Île-de-France. 2022. Dossiers: Géographie, 18 mai 2022, available online : https://www.prefectures-regions.gouv.fr/ile-de-france/Region-et-institutions/Portrait-de-la-region/Geographie/Geographie retrieved on 01.07.2024.

Lau A. 2012. Masdar City: A model of urban environmental sustainability. *Social Sciences.* https://web.stanford.edu/group/journal/cgi-bin/wordpress/wp-content/uploads/2012/09/Lau_SocSci_2012.pdf retrieved on 14.04.2014.

Lopes I.J.C., Biondi D., Corte A., Reis A., Oliveira T. 2023. A methodological framework to create an urban greenway network promoting avian connectivity: A case study of Curitiba City. *Urban Forestry & Urban Greening*, 87, 128050. https://doi.org/10.1016/j.ufug.2023.128050.

Maćków Z. 2017. Nowe Żerniki. Osiedle europejskiej stolicy kultury Wrocław 2016 [Nowe Żerniki. The Estate of the European Capital of Culture in Wrocław 2016]. *Studia KPZK*, 2017(180). https://doi.org/10.24425/118619.

Martínez C. 2015. Energy and sustainable development in cities: A case study of Bogotá. *Energy*, 92(3), 612–621. ISSN 0360-5442. https://doi.org/10.1016/j.energy.2015.02.003

Martínez J.-G., Boas I., Lenhart J., Mol P.J. 2016. Revealing Curitiba's flawed sustainability: How discourse can prevent institutional change. *Habitat International*, 53, 350–359. https://doi.org/10.1016/j.habitatint.2015.12.007

Masbuongi A. 2006. *Grand Prix de l'urbanisme 2004: Christian de Portzamparc.* Collection: Projet urbain, Parenthèses Editions, Paris.

Masdar City was named "Best Free Zone for Start Up Support" in 2017 by fDi magazine, part of the FDI intelligence services portfolio provided by the UK's Financial Times, available online: https://www.centreforpublicimpact.org/case-study/masdar-city retrieved on 07.03.2024.

Millennium Ecosystem Assessment. 2003. *Ecosystems and Human Well-Being: A Framework for Assessment.* Island Press, Washington, DC, available online: https://pdf.wri.org/ecosystems_human_wellbeing.pdf retrieved on 28.08.2024.

Mindat.org. 2024. Maclean, Clarence Valley, State of New South Wales, Australia, available online: https://www.mindat.org/feature-2159185.html, retrieved on: 12.07.2024.

Mottl T. 2021. *Narara Ecovillage Beacon for Sustainability,* Global Ecoviallage Network, published on March 29, 2016, updated on October 20, 2021, available online: https://ecovillage.org/ecovillage/narara-ecovillage/ retrieved on 29.06.2024.

MZT Aerospace Industrial Park. 2024. Brochure, available online: https://aerospacepark.mx/wp-content/uploads/2022/10/Brochure-MZT-Aerospace-v4-2022.pdf retrieved on 29.04.2024.

National Geographic Education Blog. 2007. Curitiba, Brazil: A leader in sustainable development, available online: https://blog.education.nationalgeographic.org/2007/10/02/curitiba_brazil_a_leader_in_sustainable_development/ retrieved on 30.04.2024.

Narara Ecovillage. 2024a. Available online: Nararaecovillage.com retrieved on 29.06.2024.

Narara Ecovillage. 2024b. About us, available online: https://nararaecovillage.com/about-us-2/ retrieved on 29.06.2024.

New Capital Compounds, Castle Landmark New Capital. 2024. Available online: https://www.newcapitalcompound.com/en/castle-landmark-new-capital/, retrieved on: 17.07.2024.

North Downtown Athens Redevelopment. 2024. Official webpage, available online: https://ndathens.com/ retrieved on 27.04.2024.

NSW Government. 2024. The NSW climate, available online: https://www.climatechange.environment.nsw.gov.au/basics-climate-change/nsw-climate-drivers-and-changes retrieved on 29.06.2024.

Omar W.F. 2018. Zero Carbon City – Masdar City critical analysis. *BAU Journal - Health and Wellbeing*, 1(3), Article 1. https://doi.org/10.54729/2789-8288.1053 retrieved on 11.04.2024.

O'Sullivan F., Bliss L. 12 November 2020. The 15-minute city—no cars required—is urban planning's new Utopia. *Businessweek. Bloomberg.* https://www.bloomberg.com/news/features/2020-11-12/paris-s-15-minute-city-could-be-coming-to-an-urban-area-near-you, Retrieved on 29.03.2021.

Oregon State University. 2024. Appendix C: Koppen Geiger classification descriptions, available online: https://open.oregonstate.education/permaculturedesign/back-matter/koppen-geiger-classification-descriptions/ retrieved on 01.07.2024.

OzGreen 2022. Forest Haven Eco village tour, September 2022, available online: https://www.ozgreen.org/capacity/forest-haven-eco-village-tour-2022 retrieved on 12.07.2024.

Paris-city.fr. 2024. Geographie, available online: https://www.paris-city.fr/FR/paris-city/geographie.php retrieved on 01.07.2024.

Paris Courthouse 2010–2017, Paris, France. 2024. Available online: https://www.rpbw.com/project/paris-courthouse, retrieved on 17.04.2024.

Pedraza J. 2024. From pollution to progress: Bogotá's strategy for harmonizing climate, air quality and health, SEI Published on 24 January 2024/York, UK/Bogota, Colombia, available online: https://www.sei.org/features/bogotas-climate-change-strategy-to-align-with-air-pollution-and-health-agendas-sei/ retrieved on 01.05.2024.

Perry C. 1939. Housing for the Machine Age, Russel Sage Foundation, New York, available online: https://www.russellsage.org/sites/default/files/Housing-Machine-Age.pdf retrieved on 03.04.2024.

Prestes O.M., Ultramari C., Caetano F.D. 2022. Public transport innovation and transfer of BRT ideas: Curitiba, Brazil as a reference model. *Case Studies on Transport Policy*, 10(1), 700–709. https://doi.org/10.1016/j.cstp.2022.01.031 retrieved on 03.04.2024.

Property Finder. 2024. Your guide to Masdar City Central Park, available online: https://www.propertyfinder.ae/blog/masdar-city-central-park/ retrieved on 7.03.2024.

Ramirez L.J. 2022. La Perseverancia y El Cortijo, los primeros ecobarrios de Bogotá ¡Conócelos!, article published on 6 september 2022, official webpage of Bogota, available online: https://bogota.gov.co/mi-ciudad/habitat/que-son-ecobarrios-y-cuantos-tiene-la-ciudad-de-bogota-enterate retrieved on 23.05.2024.

Reyes F. 2022. Bogotá is among the 100 most sustainable cities in the world, according to study, available online: https://bogota.gov.co/en/international/bogota-among-100-most-sustainable-cities-world-says-study retrieved on 01.05.2024.

Salingaros N.A. 2000. Complexity and urban coherence. *Journal of Urban Design*, 5, 291–316.

Salingaros, N.A. 2006. Compact city replaces sprawl. In: Graafland A., Kavanaugh L.J. (eds.) *Crossover: Architecture, Urbanism, Technology*. 010 Publishers, Rotterdam, The Netherlands, pp. 100–115.

Smith T. 2022. Cape Town scores highest of all Africal cities on measures of sustainability ESI Africa, June 21, 2022, available online: https://www.esi-africa.com/energy-efficiency/cape-town-scores-highest-of-all-african-cities-on-measures-of-sustainability/ retrieved on 13.07.2024.

Stibel E. 2022. Narara Ecovillage – Taking sustainability into their own hands, available online: https://www.stiebel-eltron.com.au/narara-ecovillage-a-community-taking-sustainability-into-their-own-hands?searchquery=35788 retrieved on 17.10.2022.

Sustainable futures Australia Our Eco-Neighbourhood. 2024. Available online: https://www.sustainablefutures.com.au/econeighbourhood retrieved on 12.07.2024.

Sustainable Transport Award, 2005. Bogotá, Colombia, available online: https://www.staward.org/past-winners/2005-bogota-colombia retrieved on 01.05.2024.

Sustainable Transport Award. 2022. Bogotá, Colombia, available online: https://www.staward.org/past-winners/2022-bogot-colombia retrieved on 01.05.2024.

System Informacji Przestrzennej Wrocławia. 2024. Charakterystyka obszaru podlegającego ocenie, available online: https://geoportal.wroclaw.pl/www/dok/ma-2013-char-obszaru.pdf retrieved on 02.05.2024.

Tibesa Bienes Raices. 2024. Parque aeroespacial será parte del megaproyecto Corredor T-MEC, available online: https://www.bienesraicestibesa.mx/noticias/parque-aeroespacial-sera-parte-del-megaproyecto-corredor-t-mec/252 retrieved on 29.04.2024.

Twardoch A. 2018. Nowe Żerniki we Wrocławiu a Aspern Seestadt w Wiedniu. Czy wrocławska realizacja nadąża za europejskimi trendami urbanistycznymi? [W:] Rembarz G. (red.), Piękno i energia. Współczesny model budowania dzielnic mieszkaniowych w Europie, Polska Akademia Nauk Komitet Przestrzennego Zagospodarowania Kraju, t. CLXXXVIII, Warszawa.

UN. 2024. Sustainable Urban Planning (Curitiba City), United Nations, Department of Economic and Social Affairs Sustainable Development, available online:https://sustainabledevelopment.un.org/index.php?page=view&type=99&nr=57& retrieved on 30.04.2024.

Visitwroclaw.eu. 2024. Pogoda, available online: https://visitwroclaw.eu/pogoda retrieved on 01.07.2024.

Weathermondo.com. 2024. The climate of Narara, available online: https://weathermondo.com/australia/narara-149669/ retrieved on 29.06.2024.

Weatherandclimate.co.uk, 2024a. The climate of Nansha, available online: https://weatherandclimate.co.uk/china/nansha-624516/ retrieved on 22.06.2024.

Weatherandclimate.co.uk. 2024b. The climate of Northeast Ithaca, available online: https://weatherandclimate.co.uk/united-states/north-carolina/northeast-ithaca-4035698/ retrieved on 01.07.2024.

Weather and Climate, Sinaloa. 2024. Mexico climate, available online: https://weatherandclimate.com/mexico/sinaloa retrieved on 01.07.2024.

Weather Spark. 2024. Climate and average weather year round in Maclean New South Wales, Australia, available online: https://weatherspark.com/y/144654/Average-Weather-in-Maclean-New-South-Wales-Australia-Year-Round, retrieved on: 12.07.2024.

World Economic Forum. 2024. Paris is planning to become a '15-minute city', video available online: https://www.weforum.org/videos/paris-is-planning-to-become-a-15-minute-city-897c12513b/ retrieved on 11.04.2024.

Zhang L., Tan P., Diehl J. 2017. A conceptual framework for studying urban green spaces effects on health. *Journal of Urban Ecology*, 3(1), 1–13. https://doi.org/10.1093/jue/jux015.

Zhang X., et al. 2020. Indoor air design parameters of air conditioners for mold-prevention and antibacterial in island residential buildings. *International Journal of Environmental Research and Public Health,* 17(19), 7316. https://doi.org/10.3390/ijerph17197316.

Summary and conclusion

The goal of this work was to promote wider use of therapeutic landscapes research evidence in eco-neighbourhood design around the globe, in well-off as well as less favoured locations. In this book, we were examining the build environment and infrastructure, trying to find possibility to counteract the tendency of cities to become unhealthy places to live. Thus, the question behind this work was: can eco-neighbourhood become health-promoting place? The objective was to demonstrate that they can and describe the methods to improve the health-promoting qualities of everyday spaces.

The therapeutic landscape research evidence provides a well-established foundation for health-promoting and sustainable neighbourhood design. The first five chapters provided evidence about various aspects of design and management of eco-neighbourhoods – human health, governance, ecology, economy, and society. The sixth, last chapter presented selected examples of operationalization of sustainable development goals in projects on all habitated continents.

Acting locally may bring a better future for our common home and for all of us. Incorporating the evidence of therapeutic qualities of landscape and aiming to promote the sustainability of urban development may result in creation of health-promoting places.

DOI: 10.4324/9781003494898-8

References

AFD (Agence Françaie de Développement). Supporting Sustainable Urban Mobility in Curitiba, available online: https://www.afd.fr/en/carte-des-projets/supporting-sustainable-urban-mobility-curitiba, retrieved on 30.04.2024.

Alexander Ch. 1977. *A Pattern Language. Towns – Buildings – Construction*, Oxford University Press, New York.

Alexander Ch. 2002. *The Nature of Order: The Process of Creating Life*, The Centre for Environmental Structure, Berkeley, CA.

Alexander M. 2023, South Africa's weather and climate, South Africa Gateway, available online: https://southafrica-info.com/land/south-africa-weather-climate/, retrieved on: 27.12.2024

Alfonzo M. 2005. To walk or not to walk? The hierarchy of walking needs. *Environment and Behavior*, 37(6), 808–836. https://doi.org/10.1177/0013916504274016

American Horticultural Therapy Association. 2024. About Therapeutic Gardens, available online: https://www.ahta.org/about-therapeutic-gardens, retrieved on: 12.09.2024

Australian Government. 2024. Narara Ecovillage Smart Grid, available online: https://arena.gov.au/projects/narara-ecovillage-smart-grid/, retrieved on 29.06.2024.

Axelson E. 2024 Cape Town national legislative capital, South Africa, Encyclopedia Britannica, available online: https://www.britannica.com/place/Cape-Town/The-people, retrieved on 27.12.2024

Balfour R., Allen J. 2014. Improving access to green spaces. Health equity briefing 8: September 2014, UCL Institute of Health Equity, Public Health England.

Bańka A. 2016. *Behavioralne podstawy projektowania architektonicznego*, Stowarzyszenie Psychologia i Architektura, Poznań.

Bańka A. 2018. *Psychologia środwiskowa jakości życia i innowacji społecznych*, Stowarzyszenie Psychologia i Architektura, Poznań.

Barrero K. 2023. Bogotá contará con dos nuevos ecobarrios en San Cristóbal y Usme Noticia de Septiembre 23, 2023, available online: https://bogota.gov.co/mi-ciudad/habitat/bogota-contara-con-dos-nuevos-ecobarrios-en-san-cristobal-y-usme, retrieved on 23.05.2024.

Bartosz D. 2024. *Podręcznik Audytora Certyfikatu Zielony Dom v.2.01*, Polskie Stowarzyszenie Budownictwa Ekologicznego.

Beaverstock J., Hubbard P., Short J.R. 2004. Getting away with it? Exposing the geographies of the super-rich. *Geoforum*, 35(4), 401–407. https://doi.org/10.1016/j.geoforum.2004.03.001

Bechtel R.B. 1997. *Environment & Behaviour: An introduction*, Sage Publications, Thousand Oaks, CA.

Bell S.L., Wheeler B.W., Phoenix C. 2017. Using geonarratives to explore the diverse temporalities of therapeutic landscapes: Perspectives from "green" and "blue" settings. *Annals of the American Association of Geographers*, 107(1), 93–108. https://doi.org/10.1080/24694452.2016.1218269

Benyus J. 2008. A good place to settle: Biommicry, biophilia, and the return of nature's inspiration to architecture. In: Kellert S., Heerwagen J., Mador M. (Eds.), *Biophilic Design: The Theory, Science, and Practice of Bringing Buildings to Life*. John Wiley & Sons, Inc, Hoboken, NJ, pp. 27–43.

Bernal C., Durosaiye I., Hadjri K., Corredor S. Duran E., Prieto A. 2022. Neglected landscapes and green infrastructure: The case of the Limas Creek in Bogotá, Colombia. *Geoforum*, 136, 194–210. https://doi.org/10.1016/j.geoforum.2022.09.010

Bickerstaff K., Walker G. 2001. Public understandings of air pollution: The 'localisation' of environmental risk. *Global Environmental Change*, 11, 133–145.

Biddulph M. 2007. *Introduction to Residential Layout*, Architectural Press, Routledge, London. https://doi.org/10.4324/9780080468617

Bjerre L., et al. 2017. On-site and laboratory evaluations of soundscape quality in recreational urban spaces. *Noise & Health*, 19(89). Published by Wolters Kluwer – Medknow, pp. 183–192.

Boeing G., et al. 2014. LEED-ND and livability revisited. *Berkeley Planning Journal*, 27, 31–55.

Bonilauri E. 2015. The Compactness Ratio or Form Factor of a Building, emupassive.com, published online on October 26, 2015, available online: https://emupassive.com/2015/10/26/the-compactness-ratio-of-a-building/, retrieved on 21.08.2024.

Bonneti M. 2010. Rompre avec l'eco-functionalisme Reseau Territorial. *Techni.Cites*, supplement au no. 194, 28–30.

Botham D. 2020. The World's Greenest Cities: Cape Town, South Africa, published January 14, 2020, available online: https://www.litterbins.co.uk/blogs/news/the-worlds-greenest-cities-cape-town-south-africa retrieved on 13.07.2024.

Boyko T., Cooper R., Dunn N. 2021. *Designing Future Cities for Wellbeing*, Routledge, London.

Bragg E. 2019. Jagera EcoCommunity, Global Ecovillages Network, available online: https://ecovillage.org/ecovillage/jagera-ecocommunity/, retrieved on 12.07.2024.

Brodzinsky S. 2014. A Tale of Two Cities: Bogotá, Americas Quarterly, January 25, 2014, available online: https://www.americasquarterly.org/fulltextarticle/a-tale-of-two-cities-bogota/, retrieved on 01.05.2024.

Brundtland G.H. 1987. Our Common Future: Report of the World Commission on Environment and Development. Geneva, UN-Dokument A/42/427, available online: https://sustainabledevelopment.un.org/content/documents/5987our-common-future.pdf, retrieved on 05.09.2024.

Cabral P. 2022. Curitiba, Brazil: A Sustainable City, CGTN, America, available online: https://america.cgtn.com/2022/10/10/curitiba-brazil-a-sustainable-city, retrieved on 11.04.2024.

Carmona M. 2021. *Public Places Urban Spaces*, Routledge, New York, Abingdon, Oxon.

Charras K. 2023. Therapeutic environments. In: Marchand D. (Ed.), *100 Key Concepts in Environmental Psychology*. Routledge, Abington, Oxon, New York, pp. 88.

China -Britain Business Focus. 2020. Nansha's Hengli: World Financial Island, published on 28.01.2024 online, https://focus.cbbc.org/nansha/amp/amp/, retrieved on 01.06.2024.

Chow A. 2022. Il est temps de penser aux villes du quart d'heure pour la santé et l'équité, Posted by NCCEH National Collaborating Centre for Environmental Health on 14.02.2022, available online: https://ccnse.ca/resources/evidence-briefs/il-est-temps-de-penser-aux-villes-du-quart-dheure-pour-la-sante-et, retrieved on 11.04.2024.

City of Ithaca. 2024a. City of Ithaca Energy Action Plan (EAP) 2012-2016 (PDF), retrieved on 27.04.2024.

City of Ithaca. 2024b. Local Action Plan: To Reduce Greenhouse Gas Emissions for City of Ithaca Government Operations (PDF), available online: https://cityofithaca.org/DocumentCenter/View/1652/Local-Action-Plan, retrieved on 27.04.2024.

City of Ithaca. 2024c. Resolution, available online: https://www.cityofithaca.org/422/City-Commitments-Resolutions, retrieved on 27.04.2024.

Clichy -Batignolles (Paris 17th). 2024. A New Urban Quality for Northwest Paris, available online: https://www.parisetmetropole-amenagement.fr/en/clichy-batignolles-paris-17th#scrollNav-4, retrieved on 17.04.2024.

Climate Data. 2024a. Athens Climate (United States of America), available online: https://en.climate-data.org/north-america/united-states-of-america/georgia/athens-1642/, retrieved on 01.07.2024.

Climate Data. 2024b. Climate: Abu Dhabi, available online: https://en.climate-data.org/asia/united-arab-emirates/abu-dhabi-2163/, retrieved on 01.07.2024.

Climate Data. 2024c. Curitiba Climate (Brazil), available online: https://en.climate-data.org/south-america/brazil/parana/curitiba-2010/, retrieved on 01.07.2024.

Climate Data. 2024d. Mazatlán Climate (Mexico), available online: https://en.climate-data.org/north-america/mexico/sinaloa/mazatlan-871696/, retrieved on 01.07.2024.

Collaborative Housing. 2024. Narara Ecoviallage, available online: https://www.collaborativehousing.org.au/stories-narara-eco-village, retrieved on 29.06.2024.

Cooper Marcus C., Sarkissian W. 1986. *Housing as if People Mattered: Site Design Guidelines for Medium-Density Family Housing*, University of California Press, Berkeley.

Cooper -Marcus C., Barnes M. 1995. *Gardens in Healthcare Facilities: Uses, Therapeutic Benefits, and Design Recommendations,* Univesity of California, Berkeley.

Cooper -Marcus C., Barnes M., 1999. *Healing Gardens, Therapheutic Benefits and Design Recommendations*, John Wiley and sons, New York. ISBN 0-471-19203-1.

Cooper -Marcus C., Sachs N. 2014. *Therapeutic Landscapes. An Evidence-Based Approach to Designing Healing Gardens and Restorative Outdoor Spaces,* John Wiley & Sons, Inc., Hoboken, NJ, pp. 14–35.

Corburn J. 2009. *Toward the Healthy City. People, Places, and the Politics of Urban Planning*, The MIT Press, Cambridge, MA, London, England.

Costa E. Silva J.A., Steffen R.E. 2019. Urban environment and psychiatric disorders: A review of the neuroscience and biology. *Metabolism*, 100S, 153940. https://doi.org/10.1016/j.metabol.2019.07.004.

Croft W., Ding L., Prasad D. 2022. Understanding decision-making in regenerative precinct developments. *Journal of Cleaner Production*, 338, 130672. ISSN 0959-6526. https://doi.org/10.1016/j.jclepro.2022.130672

Cullen G. 1961. *The Concise Townscape*, Architectural Press, Routledge, Abingdon, Oxon, New York.

Curitiba , Brazil. 2024. Curitiba's Urban Agriculture, fci FloraCulture International, available online: https://aiph.org/green-city-case-studies/curitiba-brazil-urban-agriculture/, retrieved on 30.04.2024.

Dai W. 2021. Spatial Planning and Design for Resilience: The Case of Pearl River Delta [Dissertation (TU Delft), Delft University of Technology]. https://doi.org/10.7480/abe.2021.04

Data USA. 2024. Ithaca, NY, available online: https://datausa.io/profile/geo/ithaca-ny/, retrieved on 01.07.2024.

Day R. 2006. Place and the experience of air quality. *Health & Place*, 13, 249–260. https://doi.org/10.1016/j.healthplace.2006.01.002

Declaration of Nyéléni. 2007, February 27. Nyéléni Village, Sélingué, Mali, available online: https://nyeleni.org/IMG/pdf/DeclNyeleni-en.pdf, retrieved on: 22.08.2024.

Depeau S. 2023. Mobility. In: Marchand D. (Ed.), *100 Key Concepts in Environmental Psychology*. Routledge, Abington, Oxon, New York, p.52.

Dias de Oliveira J., Biondi D., Nunho dos Reis A.R., Viezzer J. 2021. Landscape visual and sound quality influence on noise pollution propagation in urban green areas, *Revista DYNA*, 88(219), 131–138. https://doi.org/10.15446/dyna.v88n218.94724

Diehl L. 2017. Do all gardens heal the same? *CityGreen*, 14, 8–76.

Diehl L. 2023, Spring. A framework for categorizing healing gardens. *Cultivate, Florida Horticulture for Health Network*, 3(2), available online: https://www.flhhn.com/uploads/1/3/8/6/138696150/spring_2023_cultivate_flhhn.pdf, retrieved on 12.09.2024.

Dobbs C., Hernandez-Moreno A., Reyes-Paecke S., Miranda M.D. 2018. Exploring temporal dynamics of urban ecosystem services in Latin America: The case of Bogota (Colombia) and Santiago (Chile). *Ecological Indicators*, 85, 1068–1080. ISSN 1470-160X. https://doi.org/10.1016/j.ecolind.2017.11.062.

Earth haven Narara Ecovillage Masterplan. 2024. Available online: https://earth-haven.com/narara-ecovillage-masterplan/, retrieved on 11.07.2024.

Egypt Real Estate Hub. 2024. Castle Landmark – New Capital, available online: https://www.egyptrealestatehub.co.uk/castle-landmark-new-capital, retrieved on 17.07.2024.

El sol de Mazatlan. 2021. Termina la primera etapa del Parque Aeroespacial en Mazatlán, available online: https://cobertura360.mx/2021/12/11/sonora/termina-en-mazatlan-la-primera-etapa-del-parque-aeroespacial/, retrieved on 29.04.2024.

European Parliament. 2023. Circular Economy: Definition, Importance and Benefits, published: 24.05.2023, last updated: 24.05.2023 – 11, available online: https://www.europarl.europa.eu/topics/en/article/20151201STO05603/circular-economy-definition-importance-and-benefits, retrieved on 22.08.2024.

Faber Taylor A, Kuo F.E., Sullivan W.C. 2002. Views of nature and self-discipline: Evidence from inner city children. *Journal of Environmental Psychology*, 22, 49–64.

Faber Taylor A., Kuo F.E. 2009. Children with attention deficits concentrate better after walk in the park. *Journal of Attention Disorders*, 12, 402–409.

Families on the Coast. 2021. Narara Ecovillage: Where Families Thrive in Community, sponsored editorial published on June 1, 2021, available online: https://onthecoastpublications.com.au/onthecoastfamilies/narara-ecovillage-where-families-thrive-in-community/, retrieved on 11.07.2024.

Feddema J.-J. 2005. A revised thornthwaite-type global climate classification. *Physical Geography*, 26(6), 442–466.

Feliot -Rippeault M. 2023. Biodiversity. In: Marchand D. (Ed.), *100 Key Concepts in Environmental Psychology*. Routledge, Abington, Oxon, New York, p.16.

Foo A.F. 1999. The seven lamps of sustainable city. In: Foo A.F., Yuen B. (Eds.), *Sustainable Cities in the 21st Century*. Faculty of Architecture, Building & Real Estate National University of Singapore, Singapore, pp. 109–130.

Francis , The Holy Father. 2015. Encyclical Letter *LAUDATO SI'* of the Holy Father Francis on Care for Our Common Home, available online: https://www.vatican.va/content/francesco/en/encyclicals/documents/papa-francesco_20150524_enciclica-laudato-si.html, retrieved on 11.9.2024.

Freudenberg N. 2000. Time for a national agenda to improve the health of urban populations. *American Journal of Public Health*, 90(6), 837–840. https://doi.org/10.2105/ajph.90.6.837. PMID: 10846496; PMCID: PMC1446275.

Friedman A. 2022. *Designing Innovative Sustainable Neighborhoods* (1st ed.), Routledge, London. https://doi.org/10.4324/9781003203025

Fuller R.A., Irvine K.N., Devine-Wright P., Warren P.H., Gaston K.J. 2007. Psychological benefits of greenspace increase with biodiversity. *Biology Letters*, 3(4), 390–394. https://doi.org/10.1098/rsbl.2007.0149. PMID: 17504734; PMCID: PMC2390667.

Garvin E., Cannuscio C., Branas C. 2012. Greening vacant lots to reduce violent crime: A randomised controlled trial. *Injury Prevention.* https://doi.org/10.1136/injuryprev-2012-040439

Gehl J. 2011. *Life between Buildings: Using Public Space* (6th ed.). Island Press, Washington.

Gerlach -Springs N., Kaufman R.E., Warner S.B. 1998. *Restorative Gardens.* Yale University Press, New Haven, CT.

Gerlach -Springs N., Healy V. 2010, Spring. *The Therapeutic Garden: A Definition.* Healthcare and *Therapeutic Design Newsletter.* American Society of Landscape Architecture (ASLA), Washington, DC.

Gesler W. 1996. Lourdes: Healing in a place of pilgrimage. *Health & Place*, 2(2), 95–105.

Gesler W. 2005. Therapeutic landscapes: An evolving theme. *Health & Place*, 11, 295–297.

Ghosh S. 2024. Water management practices in master-planned developments and eco-neighborhoods. In: Abraham M.A. (Ed.), *Encyclopedia of Sustainable Technologies* (2nd ed.), Elsevier, Amsterdam, Oxford, Cambridge pp. 803–819. ISBN 9780443222870, https://doi.org/10.1016/B978-0-323-90386-8.00074-7.

Global BRT Data. 2024. Available online: https://brtdata.org/#/location, retrieved on 30.04.2024.

GDCI (Global Design Cities Initiative). 2024. *Global Street Design Guide*, National Association of City Transportation Officials, New York, available online: https://globaldesign ingcities.org/publication/global-street-design-guide/resources/references, retrieved on 19.08.2024.

Gnatek T. 2003. Brazil-Curitiba's Urban Experiment, December 2003, Frontline World, available online: https://www.pbs.org/ frontlineworld/fellows/brazil1203/, retrieved on 30.04.2024.

Gobierno de Mexico, Mazatlán. 2024. Available online: https://www.economia.gob.mx/datamexico/en/profile/geo/mazatlan, retrieved on 01.07.2024.

Gorman R. 2017. Smelling therapeutic landscapes: Embodied encounters within spaces of care farming. *Health & Place*, 47, 22–28. https://doi.org/10.1016/j.healthplace.2017.06.005

GPA (The Georgia Planning Association, Official Chapter of the American Planning Association (APA)). 2024. *2023 GPA Fall Chapter Awards*, available online: https://georgiaplanning.org/wp-content/uploads/GPA-Awards-2023.pdf, retrieved on 27.04.2024.

Grave R. 2021. Parque Aeroespacial y MLC fortalecerán al sector educativo. Se acabó la fuga de cerebros en el Puerto, Punto MX, 1.12.2021, available online: https://punto.mx/2021/12/01/parque-aeroespacial-y-mlc-fortaleceran-al-sector-educativo/, retrieved on 29.04.2024.

Griffiths S., Sovacol B.K. 2020, April. Rethinking the future low-carbon city: Carbon neutrality, green design, and sustainability tensions in the making of Masdar City. *Energy Research & Social Science*, 62, 101368. https://doi.org/10.1016/j.erss.2019.101368

Grylls T., von Reeuwijk M. 2022. How trees affect urban air quality: It depends on the source. *Atmospheric Environment*, 290, 119275. https://doi.org/10.1016/j.atmosenv.2022.119275

Guislain M. 2016. Le macrolot, paysage urbain du XXI siécle? *AMC Le Moniteur architecture*, 2016, 855.

Guislain M. 2019. Christian de Portzamparc, en noir & blanc et en lumière. *Le Moniteur*, 30 decembre 2019, available online: https://www.lemoniteur.fr/article/christian-de-portzamparc-en-noir-blanc-et-en-lumiere.2069424, retrieved on 15.09.2023.

Hall P., Tewdwr-Jones M. 2020. *Urban and Regional Planning*, Routledge, Abingdon, Oxon, New York.

Hartig T., Mitchell R., de Vries S., Frumkin H. 2014. Nature and health. *Annual Review of Public Health*, 35, 207–228. https://doi.org/10.1146/annurev-publhealth-032013-182443

Harvard Energy & Facilities. 2024. Integrated Design, available online: https://www.energyandfacilities.harvard.edu/green-building-resource/green-building-tools-resources/integrated-design, retrieved on 18.07.2024.

Heffernan E., de Wilde P. 2020. Group self-build housing: A bottom-up approach to environmentally and socially sustainable housing. *Journal of Cleaner Production*, 243, 118657. ISSN 0959-6526. https://doi.org/10.1016/j.jclepro.2019.118657

Hidalgo D. 2014. *Urbanism Hall of Fame: Jaime Lerner – The Architect of Curitiba*, available online: https://thecityfix.com/blog/urbanism- hall-fame-jaime-lerner-architect-curitiba-dario-hidalgo/, retrieved on 24.02.2015.

Holgate C. 2007. Factors and actors in climate change mitigation: A tale of two South African cities. *Local Environment*, 12(5), 471–484. https://doi.org/10.1080/13549830701656994

Howden -Chapman F., Baker M., Bierre S. 2013. The houses children live in: Policies to improve housing quality. *Policy Quarterly*, 9(2), 35. https://doi.org/10.26686/pq.v9i2.4450

Huang L., Xu H. 2017. Therapeutic landscapes and longevity: Wellness tourism in Bama. *Social Science & Medicine*, 197, 24–32. https://doi.org/10.1016/j.socscimed.2017.11.052

Ignatieva M. 2012. Plant material for urban landscapes in the era of globalization: Roots, challenges and innovative solutions. In: Richter M., Weiland U. (Eds.), *Applied Urban Ecology: A Global Framework*. Wiley-Blackwell, Chichester, UK, Hoboken, NJ, pp. 139–151. https://doi.org/10.1002/9781444345025.ch11.

Industrial Park MZT Aerospace. 2024. Available online: https://aerospacepark.mx/, retrieved on 29.04.2024.

Invest Cape Town. 2024. Cape Town Makes Serious Strides Towards a Green Economy, available online: https://www.investcapetown.com/why-cape-town/green-economy/, retrieved on 13.07.2024.

IPHA (International Passive House Association). What Is a Passive House? Passipedia, available online: https://passipedia.org/basics/what_is_a_passive_house, retrieved on 21.08.2021.

ITDP (Institute for Transportation and Development Policy). 2009. Lessons of Urban Transit from Bogota, available online: https://www.itdp.org/lessons-of-urban-transit-from-bogota/, retrieved on 27.05.2024.

Jackobs J. 1961. *The Death and Life of Great American Cities*, Random House, New York City.

Jagera EcoCommunity. 2019. Application for membership of global ecovillage network Australia, available online: https://static1.squarespace.com/static/5c367f6ec3c16a98d012bb9c/t/5c528151c74c50734e5e5137/1548910933981/Jagera+Ecovillage+Accreditation+15+Jan+2019.pdf retrieved on 12.07.2024.

Jagera EcoCommunity. 2023. Available online: https://www.ecocommunity.org.au/ retrieved on 12.07.2024.

James P. 2018. *The Biology of Urban Environments*, Oxford University Press, Oxford. © Philip James 2018. https://doi.org/10.1093/oso/9780198827238.001.0001

Jan Paweł II, papież. 1987. Encyklika Sollicitudo Rei Socialis. Watykan (SRS 27–28), available online: https://adonai.pl/jp2/pliki/srs.pdf, retrieved on 10.06.2024.

Jara -Moreno A. 2010. TransMilenio Celebrates 10 Years of Operation, article published on 03.11.2010, available online: https://www.itdp.org/transmilenio-celebrates-10-years-of-operation/, retrieved on 27.05.2024.

Jatzlauk G., Bartel S., Heine H., Schloter M., Krauss-Etschmann S. 2017. Influences of environmental bacteria and their metabolites on allergies, asthma, and host microbiota. *Allergy*, 72(12), 1859–1867. https://doi.org/10.1111/all.13220. Epub 2017 June 28. PMID: 28600901.

Jenkins P., Smith H. 2024. *Order and Disorder in Urban Space and Form*, Routledge, Abingdon, Oxon, New York. ? URL is valid

JHP. 2021. *North Downtown Athens Master Planning Study Redevelopment Masterplan for Bethel Village/ College & Hoyt*, Athens, March 2, 2021, available online: h https://ndathens.com/wp-content/uploads/2022/03/2019069_210301_Feb-final-report_Small.pdf, retrieved on 27.04.2024.

JHP Architecture/Urban Design. 2023. North Downtown Athens Master Plan Wins GPA Award, available online: https://jhparch.com/north-downtown-athens-master-plan-wins-gpa-award, retrieved on 27.04.2024.

Kahn P.H. 1999. *The Human Relationship with Nature. Development and Culture*, The MIT Press, Cambridge, MA, London, England.

Kaplan S. 1995. The restorative benefits of nature – Towards an integrative framework. *Journal of Environmental Psychology*, 15(3), 169–182.

Kaplan R., Kaplan S., Ryan R.L. 1998. *With People in Mind. Design and Management of Everyday Nature*, Island Press, Washington, DC, Covelo, CA.

Kapoor M.K. 2023. *Security by Design: Protecting Buildings and Public Places against Crime and Terror*, Routledge, London.

Karmanov D., Hamel R. 2008. Assessing the restorative potential of contemporary urban environment(s): Beyond the nature versus urban dichotomy. *Landscape and Urban Planning*, 86, 115–125.

Kavanagh J.S. 2011. A Thumbnail History of Therapeutic Gardens in Healthcare, LATIS 2011, Landscape Architecture Technical Information Series Number 1, Forum on Therapeutic Garden Design, 2nd ed., available online: https://www.asla.org/uploadedFiles/CMS/Practice/Research_Reports/ASLA_Research_TherapeuticGardenDesign_2011.pdf, retrieved on 12.09.2024.

Kobylarczyk J. 2018. Uwarunkowania środowiskowe w projektowaniu obszarów mieszkaniowych, Wydawnictwo Politechniki Krakowskiej, Kraków.

Konijnendijk C.C. 2023. Evidence-based guidelines for greener, healthier, more resilient neighbourhoods: Introducing the 3–30–300 rule. *Journal of Forestry Research,* 34, 821–830. https://doi.org/10.1007/s11676-022-01523-z

Kottek M., Grieser J., Beck C., Rudolf B., Rubel F. 2006. World map of the Köppen-Geiger climate classification updated. *Meteorologische Zeitschrif,* 15(3), 259–263. https://doi.org/10.1127/0941-2948/2006/0130

Krier L. 2011. Architektura wspólnoty, Wydawnictwo: słowo/obraz/terytoria, Gdańsk.

Kuo F. 2010. *Parks and Other Green Environments: Essential Components of a Healthy Human Habitat*, National Recreation and Park Association, Asburn, VA.

Kuo F., Sullivan W. 2001. Environment and crime in the inner city. Does vegetation reduce crime? *Environment and Behavior*, 33(3), 343–367.

Kuo F., Bacaicoa M., Sullivan W. 1998. Transforming inner-city landscapes. Trees, sense of safety, and preference. *Environment and Behaviour*, 30(1), 28–59.

La préfecture et les services de l'Etat en région l'Île-de-France. 2022. Dossiers: Géographie, 18 mai 2022, available online: https://www.prefectures-regions.gouv.fr/ile-de-france/Region-et-institutions/Portrait-de-la-region/Geographie/Geographie, retrieved on 01.07.2024.

Lang J. 2017. *Urban Design*, Routledge, Abingdon, Oxon, New York.

Lau A. 2012. Masdar City: A Model of Urban Environmental Sustainability, Social Sciences, available online: https://web.stanford.edu/group/journal/cgi-bin/wordpress/wp-content/uploads/2012/09/Lau_SocSci_2012.pdf, retrieved on 14.04.2014.

Lawrence R.J. 2020. *Creating Built Environments: Bridging Knowledge and Practice Divides*, Routledge, Abingdon, Oxon, New York.

Lederbogen F., et al. 2011. City living and urban upbringing affect neural social stress processing in humans. *Nature*, 474(7352), 498–501. https://doi.org/10.1038/nature10190

Lima M.-L. 2023. Landscape. In: Marchand D. (Ed.), *100 Key Concepts in Environmental Psychology*. Routledge, Abington, Oxon, New York, p. 47.

Lopes I.J.C., Biondi D., Corte A., Reis A., Oliveira T. 2023. A methodological framework to create an urban greenway network promoting avian connectivity: A case study of Curitiba City. *Urban Forestry & Urban Greening*, 87, 128050. https://doi.org/10.1016/j.ufug.2023.128050

Lothian A. 2017. *The Science of Scenery: How We See Scenic Beauty, What It Is, Why We Love It, and How to Measure and Map*, CreateSpace Independent Publishing Platform, Scotts Valley, CA.

Louv R. 2005. *Last Child in the Woods: Saving Our Children from Nature-Deficit Disorder*, Algonquin Books, New York.

Lovasi G.S., Quinn J.W., Neckerman K.M., Perzanowski M.S., Rundle A. 2008. Children living in areas with more street trees have lower prevalence of asthma. *Journal of Epidemiology & Community Health*, 62(7), 647.

Lynch K. 1961. *The Image of the City*, The MIT Press, Cambridge, MA, London, England.

Maas J., Verheij R.A., de Vries S., Spreeuwenberg P., Schellevis F., Groenewegen P.P. 2006. Green space, urbanity, and health: How strong is the relation? *Journal of Epidemiology and Community Health*, 60(7), 587–592. DOI: 10.1136/jech.2005.043125

Maas J. 2008. *Vitamin G: Green Environments – Healthy Environments*. Nivel, Ultrecht.

Mass J., van Dillen S.M.E., Verheij R.A., Groenewegen R.P. 2009. Social contacts as possible mechanizm behind the relation between Green space and health. *Health and Place*, 15(2), 586–595.

Maas J., Verheij R.A. 2009. Morbidity is related to a Green living environment. *Journal of Epidemiology and Community Health*, 63, 967–973.

Maćków Z. 2017. Nowe Żerniki. Osiedle europejskiej stolicy kultury Wrocław 2016 [Nowe Żerniki. The Estate of the European Capital of Culture in Wrocław 2016]. *Studia KPZK*, (180). https://doi.org/10.24425/118619.

Marchand D. 2023. *100 Key Concepts in Environmental Psychology*, Routledge, Abington, Oxon, New York.

Marsh E., et al. 2022. Endogenous mood state and hedonic responses to pleasant odors. *Journal of Sensory Studies*, 38, e12826. https://doi.org/10.1111/joss.12826.

Martínez C. 2015. Energy and sustainable development in cities: A case study of Bogotá. *Energy*, 92(3), 612–621. ISSN 0360-5442. https://doi.org/10.1016/j.energy.2015.02.003

Martínez J.-G., Boas I., Lenhart J., Mol P.J. 2016. Revealing Curitiba's flawed sustainability: How discourse can prevent institutional change. *Habitat International*, 53, 350–359. https://doi.org/10.1016/j.habitatint.2015.12.007

Masbuongi A. 2006. Grand Prix de l'urbanisme 2004: Christian de Portzamparc, Collection: Projet urbain, Parenthèses Editions, Paris.

Masdar City was named "Best Free Zone for Start Up Support" in 2017 by fDi magazine, part of the FDI intelligence services portfolio provided by the UK's Financial Times, available online: https://www.centreforpublicimpact.org/case-study/masdar-city, retrieved on 7.03.2024.

Maslow A.H. 1954. *Motivation and Personality*, Harper & Brothers, New York.

McPherson E.G., Nowak D.J., Rowntree R.A. 1994. Chicago's Urban Forest Ecosystem: Results of the Chicago Urban Forest Climate Project. USDA Forest Service General Technical Report NE-186.

Millennium Ecosystem Assessment. 2003. *Ecosystems and Human Well-being: A Framework for Assessment*, Island Press, Washington, DC, available online: http://pdf.wri.org/ecosystems_human_wellbeing.pdf, retrieved on 28.08.2024.

Milligan C., Gatrell A., Bingley A. 2004. 'Cultivating health': Therapeutic landscapes and older people in northern England. *Social Science & Medicine*, 58, 1781–1793.

Milman O. 2019, September 20. US to stage its largest ever climate strike: 'Somebody must sound the alarm', available online: www.theguardian.com/world/2019/sep/20/climate-strikes-us-students-greta-thunberg, retrieved on 29.07.2020.

Mindat. org. 2024. Maclean, Clarence Valley, State of New South Wales, Australia, available online: https://www.mindat.org/feature-2159185.html, retrieved on 12.07.2024.

Minh A., Muhajarine N., Janus M., Brownell M., Guhn M. 2017. A review of neighborhood effects and early child development: How, where, and for whom, do neighborhoods matter? *Health Place*, 46, 155–174. https://doi.org/10.1016/j.healthplace.2017.04.012. Epub 2017 May 18. PMID: 28528276.

Mitchell R., Popham F. 2008. Effect of exposure to natural environment on health inequalities: An observational population study. *Lancet*, 372(9650), 1655–1660. DOI: 10.1016/S0140-6736(08)61689-X

Montgomery Ch. 2015. *Happy City: Transforming Our Lives Through Urban Design*, Penguin, London.

Mottl T. 2021. *Narara Ecovillage Beacon for Sustainability*, Global Ecoviallage Network, published on March 29, 2016, updated on October 20, 2021, available online: https://ecovillage.org/ecovillage/narara-ecovillage/, retrieved on 29.06.2024.

Moreno C., Allam Z., Chabaud D., Gall C., Pratlong F. 2021. Introducing the '15-Minute City': Sustainability, resilience and place identity in future post-pandemic cities. *Smart Cities*, 4(1), 93–111. https://doi.org/10.3390/smartcities4010006

Moudon A.V., Lee Ch., Cheadle A.D., Garvin C., Johnson D., Schmid T.L., Weathers R.D., Lin L. 2006. Operational definitions of walkable neighborhood: Theoretical and empirical insights. *Journal of Physical Activity and Health*, 3(Suppl 1), S99–S117.

Musy M. (Ed.). 2014. Une ville vert. Les rôles du végétal en ville. Éditions Quæ, Versaille.

MZT Aerospace Industrial Park brochure. 2024. Available online: https://aerospacepark.mx/wp-content/uploads/2022/10/Brochure-MZT-Aerospace-v4-2022.pdf, retrieved on 29.04.2024.

Nara Document on Authenticity, UNESCO. 1994. Available online: https://www.icomos.org/en/charters-and-texts/179-articles-en-francais/ressources/charters-and-standards/386-the-nara-document-on-authenticity-1994, retrieved on 28.08.2024.

Narara Ecovillage. 2024. About us, available online: https://nararaecovillage.com/about-us-2/, retrieved on 29.06.2024.

National Geographic Education Blog. 2007. Curitiba, Brazil: A Leader in Sustainable Development, available online: https://blog.education.nationalgeographic.org/2007/10/02/curitiba_brazil_a_leader_in_sustainable_development/, retrieved on 30.04.2024.

Navarro O. 2023. Connection with nature. In: Marchand D. (Ed.), *100 Key Concepts in Environmental Psychology*. Routledge, Abington, Oxon, New York, p. 19.

New Capital Compounds, Castle Landmark New Capital. 2024. Available online: https://www.newcapitalcompound.com/en/castle-landmark-new-capital/, retrieved on: 17.07.2024.

North Downtown Athens Redevelopment official webpage. 2024. Available online: https://ndathens.com/, retrieved on 27.04.2024.

Nowak D.J., Crane D.E. 2000. The Urban Forest Effects (UFORE) model: Quantifying urban forest structure and functions. In: Hansen M., Burk T. (Eds.), *Integrated Tools for Natural Resources Inventories in the 21st Century. Proceedings of the IUFRO Conference*. USDA Forest Service General Technical Report NC-212, pp. 714e720.

Nowak D.J., Crane E.E., Stevens J.C., Ibarra M. 2002. Brooklyn's Urban Forest. USDA Forest Service General Technical Report NE-290.

NSW Government. 2024. The NSW Climate, available online: https://www.climatechange.environment.nsw.gov.au/basics-climate-change/nsw-climate-drivers-and-changes, retrieved on 29.06.2024.

Oldenburg R. 1989. *The Great Good Place*. Da Capo Press, Boston, MA.

Omar W.F. 2018. Zero carbon city – Masdar city critical analysis. *BAU Journal – Health and Wellbeing*, 1(3), Article 1. https://digitalcommons.bau.edu.lb/cgi/viewcontent.cgi?params=/context/hwbjournal/article/1053/&path_info=1_ZERO_CARBON_CITY__MASDAR_CITY_CRITICAL_ANALYSIS.pdf, retrieved on 11.04.2024.

Oregon State University. 2024. Appendix C: Koppen Geiger Classification Descriptions, available online: https://open.oregonstate.education/permaculturedesign/back-matter/koppen-geiger-classification-descriptions/, retrieved on 01.07.2024.

O'Sullivan F., Bliss L. 2020, November 12. The 15-Minute City—No Cars Required—Is Urban Planning's New Utopia. Businessweek. Bloomberg, retrieved on 29.03.2021.

Ovans A. 2015. *What Resilience Means and Why It Matters*. Harvard Business Review.

OzGreen. 2022. Forest Haven Eco Village Tour, September 2022, available online: https://www.ozgreen.org/capacity/forest-haven-eco-village-tour-2022, retrieved on 12.07.2024.

Paris-city.fr. 2024. Geographie, available online: http://www.paris-city.fr/GB/paris-city/geographie.php, retrieved on 01.07.2024.

Paris Courthouse. 2010–2017. Paris, France, available online: https://www.rpbw.com/project/paris-courthouse, retrieved on 17.04.2024.

Pedraza J. 2024. From pollution to progress: Bogotá's strategy for harmonizing climate, air quality and health, SEI Published on 24 January 2024, York/Bogota, Colombia, available online: https://www.sei.org/features/bogotas-climate-change-strategy-to-align-with-air-pollution-and-health-agendas-sei/, retrieved on 01.05.2024.

Perry C. 1939. Housing for the Machine Age, Russel Sage Foundation, New York, available online: https://www.russellsage.org/sites/default/files/Housing-Machine-Age.pdf, retrieved on 03.04.2024.

Picon A. 2015. *Smart Cities: A Spatialised Intelligence*, Wiley, New York.

Porteous , J. D. (1985). Smellscape. *Progress in Physical Geography: Earth and Environment*, *9*(3), 356-378. https://doi.org/10.1177/030913338500900303

Prestes O.M., Ultramari C., Caetano F.D. 2022. Public transport innovation and transfer of BRT ideas: Curitiba, Brazil as a reference model. *Case Studies on Transport Policy*, 10(1), 700–709. https://doi.org/10.1016/j.cstp.2022.01.031, retrieved on 03.04.2024.

Prior J. 2017. Sonic environmental aesthetics and landscape research. *Landscape Research*, 42(1), 6–17. https://doi.org/10.1080/01426397.2016.1243235

Ramirez L.J. 2022. La Perseverancia y El Cortijo, los primeros ecobarrios de Bogotá ¡Conócelos! , article published on 6 September 2022, official webpage of Bogota, available online: https://bogota.gov.co/mi-ciudad/habitat/que-son-ecobarrios-y-cuantos-tiene-la-ciudad-de-bogota-enterate, retrieved on 23.05.2024.

Rayner G., Lang T. 2012. *Ecological Public Health. Reshaping the Conditions for Good Health*, Routledge, Abingdon and New York.

Reyes F. 2022. Bogotá is among the 100 most sustainable cities in the world, according to study, available online: https://bogota.gov.co/en/international/bogota-among-100-most-sustainable-cities-world-says-study, retrieved on 01.05.2024.

Richard I. 2023. Renaturation. In: Marchand D. (Ed.), *100 Key Concepts in Environmental Psychology*. Routledge, Abington, Oxon, New York, p.110.

Roe J., McCay L. 2021. *Restorative Cities: Urban Design for Mental Health and Wellbeing*, Bloomsbury Publishing, London.

Roussel J. 2023. Walkability. In: Marchand D. (Ed.), *100 Key Concepts in Environmental Psychology*. Routledge, Abington, Oxon, New York, p.165.

Sæbø A., Popek R., Nawrot B., Hanslin H.M., Gawronska H., Gawronski S.W. 2012. Plant species differences in particulate matter accumulation on leaf surfaces. *Science of the Total Environment*, 427–428, 347–354. ISSN 0048-9697. https://doi.org/10.1016/j.scitotenv.2012.03.084

Salingaros N.A. 2000. Complexity and urban coherence. *Journal of Urban Design*, 5, 291–316.

Salingaros, N.A. 2006. Compact city replaces sprawl. In: Graafland A., Kavanaugh L.J. (ed.) *Crossover: Architecture, Urbanism, Technology*. 010 Publishers, Rotterdam, The Netherlands, pp. 100–115.

Serfaty -Garzon P. 2023a. Home (chez-soi). In: Marchand D. (Ed.), *100 Key Concepts in Environmental Psychology*. Routledge, Abington, Oxon, New York.

Serfaty -Garzon P. 2023b. Housing. In: Marchand D. (Ed.), *100 Key Concepts in Environmental Psychology*. Routledge, Abington, Oxon, New York, p.74.

Sharifi A., Murayama A. 2014. Neighborhood sustainability assessment in action: Cross-evaluation of three assessment systems and their cases from the US, the UK, and Japan. *Building and Environment*, 72, 243–258. https://doi.org/10.1016/j.buildenv.2013.11.006

Sharifi A., Murayama A. 2015. Viability of using global standards for neighbourhood sustainability assessment: Insights from a comparative case study. *Journal of Environmental Planning and Management*, 58(1), 1–23. https://doi.org/10.1080/09640568.2013.866077

Simoes Aelbrecht P. 2016. 'Fourth places': The contemporary public settings for informal social interaction among strangers. *Journal of Urban Design*, 21(1), 124–152. https://doi.org/10.1080/13574809.2015.1106920

Smith T. 2022. Cape Town Scores Highest of All Africal Cities on Measures of Sustainability ESI Africa, June 21, 2022, available online: https://www.esi-africa.com/energy-efficiency/cape-town-scores-highest-of-all-african-cities-on-measures-of-sustainability/, retrieved on 13.07.2024.

Staats H. 2023. Nature, a psychological perspective. In: Marchand D. (Ed.), *100 Key Concepts in Environmental Psychology*. Routledge, Abington, Oxon, New York, pp. 87–90.

Sternberg E. 2010. *Healing Spaces. The Science of Place and Well-Being*, The Belknap Press of Harvard University Press, Cambridge, London.

Stibel E. 2022. Narara Ecovillage – Taking Sustainability into Their Own Hands, available online: https://www.stiebel-eltron.com.au/narara-ecovillage-a-community-taking-sustainability-into-their-own-hands?searchquery=35788, retrieved on 17.10.2022.

Stigsdotter U.A., Grahn P. 2002. What makes a garden a healing garden? *Journal of Therapheutic Horticulture*, 13, 60–69.

Sustainable Futures Australia Our Eco-Neighbourhood. 2024. Available online: https://www.sustainablefutures.com.au/econeighbourhood, retrieved on 12.07.2024.

Sustainable Transport Award. 2005. Bogotá, Colombia, available online: https://www.staward.org/past-winners/2005-bogota-colombia, retrieved on 01.05.2024.

Sustainable Transport Award. 2022. Bogotá, Colombia, available online: https://www.staward.org/past-winners/2022-bogot-colombia, retrieved on 01.05.2024.

System Informacji Przestrzennej Wrocławia. 2024. Charakterystyka obszaru podlegającego ocenie, available online: https://geoportal.wroclaw.pl/www/dok/ma-2013-char-obszaru.pdf, retrieved on 02.05.2024.

Takano T ., Nakamura K., Watanabe M. 2002. Urban residential environments and senior citizens' longevity in megacity areas: The importance of walkable green spacer. *Journal of Epidemiology Community Health*, 56, 913–918.

Tallis M ., Taylor G., Sinnett D., Freer-Smith P. 2011. Estimating the removal of atmospheric particulate pollution by the urban tree canopy of London, under current and future environments. *Landscape and Urban Planning*, 103(2), 129–138.

Tibesa B ienes Raices. 2024. Parque aeroespacial será parte del megaproyecto Corredor T-MEC, available online: https://www.bienesraicestibesa.mx/noticias/parque-aeroespacial-sera-parte-del-megaproyecto-corredor-t-mec/252, retrieved on 29.04.2024.

Trojanowska M . 2017. *Parki i ogrody terapeutyczne*, Wydawnictwo Naukowe PWN, Warszawa.

Trojanowska M., Sas-Bojarska A. 2018. *Health-affirming everyday landscapes in sustainable city. Theories and tools* Architecture Civil Engineering Environment Journal, ACEE 2018, 11, 3, 41–52.

Trojanowska M . 2020. *Poszukiwanie standard projektowania ekoosieldi w Polsce*, Wydawnictwa Naukowe Uniwersytetu Technologiczno-Przyrodniczego w Bydgoszczy, Bydgoszcz.

Trojanowska M . 2022. Parki i ogrody terapeutyczne w służbie zdrowia psychicznego. In: Gawrych M. (Ed.), *Natura a zdrowie psychiczne, Wydawnictwo Akademii Pedagogiki Specjalnej im.* Marii Grzegorzewskiej, Warszawa, pp. 151–177

Trojanowska M . 2023a. *Projektowanie zielonych przestrzeni publicznych*, Wydawnictwo Naukowe PWN, Warszawa.

Trojanowska M . 2023b. Reclamation of polluted land in urban renewal projects. Literature review of suitable plants for phytoremediation. *Environmental Challenges*, 13, 1–7, 100749. https://doi.org/10.1016/j.envc.2023.100749

Trojanowska M . 2024. The evolving theme of health-promoting urban form: Applying the macrolot concept for easy access to open public green spaces. *Urban Science*, 8(3), 115. https://doi.org/10.3390/urbansci8030115

Twardoch A . 2018. Nowe Żerniki we Wrocławiu a Aspern Seestadt w Wiedniu. Czy wrocławska realizacja nadąża za europejskimi trendami urbanistycznymi? In: Rembarz G. (red.), Piękno i energia. Współczesny model budowania dzielnic mieszkaniowych w Europie, Polska Akademia Nauk Komitet Przestrzennego Zagospodarowania Kraju, t. CLXXXVIII, Warszawa, pp. 174–192.

Ulrich R.S. 1984. View through a window may influence recovery from surgery. *Science*, 224(4647), 420–421. https://doi.org/10.1126/science.6143402. PMID: 6143402.

Ulrich , R. S. (1999) Effects of gardens on health outcomes: theory and research. In: Cooper Marcus C. and Barnes M. (1999). *Healing Gardens: Therapeutic Benefits and Design.* Wiley (Series in Healthcare and Senior Living Design): New York, p. 624.

Urlich R. 2023. Stress reduction theory. In: Marchand D. (Ed.), *100 Key Concepts in Environmental Psychology*. Routledge, Abington, Oxon, New York, pp.143–145.

UN. 2024. Sustainable Urban Planning (Curitiba City), United Nations, Department of Economic and Social Affairs Sustainable Development, available online: https://sustainabledevelopment.un.org/index.php?page=view&type=99&nr=57&, retrieved on 30.04.2024.

Valera S. 2023. Wayfinding. In: Marchand D. (Ed.), *100 Key Concepts in Environmental Psychology*. Routledge, Abington, Oxon, New York, pp.166.

Velarde M.D., Fry G., Tveit M. 2007. Health effects of viewing landscapes – Landscape types in environmental psychology. *Urban Forestry & Urban Greening*, 6, 199–212.

Vidal T. 2023. Eco-district. In: Marchand D. (Ed.), *100 Key Concepts in Environmental Psychology*. Routledge, Abington, Oxon, New York, pp.35.

Visitwroclaw. eu. 2024. Pogoda, available online: https://visitwroclaw.eu/pogoda, retrieved on 01.07.2024.

Vos P., Maiheu B., Vankerkom J., Janssen S. 2013. Improving local air quality in cities: To tree or not to tree? *Environmental Pollution*, 183, 113–122. ISSN 0269-7491. https://doi.org/10.1016/j.envpol.2012.10.021

Walker. 2024. Design for the Other 90%, available online: https://walkerart.org/calendar/2008/design-for-the-other-90/, retrieved on 04.09.2024.

Ward Thompson C., Aspinall P., Bell S. 2010. *Innovative Approaches to Researching Landscape and Health*. Open Space: People Space 2. Routledge, Abingdon and New York.

Weathermondo. com. 2024. The climate of Narara, available online: https://weathermondo.com/australia/narara-149669/, retrieved on 29.06.2024.

Weatherandclimate. co.uk. 2024. The climate of Northeast Ithaca, available online: https://weatherandclimate.co.uk/united-states/north-carolina/northeast-ithaca-4035698/, retrieved on 01.07.2024.

Weather and Climate. 2024. Sinaloa, Mexico Climate, available online: https://weatherandclimate.com/mexico/sinaloa, retrieved on 01.07.2024.

Weatherandclimate. co.uk. 2024. The Climate of Nansha, available online: https://weatherandclimate.co.uk/china/nansha-624516/, retrieved on 22.06.2024.

Weather Spark. 2024. Climate and Average Weather Year Round in Maclean New South Wales, Australia, available online: https://weatherspark.com/y/144654/Average-Weather-in-Maclean-New-South-Wales-Australia-Year-Round, retrieved on 12.07.2024.

Weiss K. 2023. Place Attachement, Place attachment refers to the affective and cognitive bond that individuals develop with a place.

World Health Organization (WHO). Regional Office for Europe. (1989). European Charter on Environment and Health, 1989. World Health Organization. Regional Office for Europe. https://iris.who.int/handle/10665/347390

WHO. 2010. Global recommendations on physical activity for health, available online at: https://www.who.int/publications/i/item/9789241599979, retrieved on 07.09.2024.

WHO. 2024. About the Global Network for Age-friendly Cities and Communities, available online: https://extranet.who.int/agefriendlyworld/who-network/, retrieved on 04.09.2024.

WHO Healthy Cities. 2024. Available online: https://www.who.int/southeastasia/activities/healthy-cities, retrieved on 05.09.2024.

WHO Regional Office for Europe. 2016. Urban green spaces and health, available online: https://iris.who.int/bitstream/handle/10665/345751/WHO-EURO-2016-3352-43111-60341-eng.pdf?sequence=3&isAllowed=y, retrieved on 13.08.2024.

Williams A. 1998. Therapeutic landscapes in holistic medicine. *Social Science and Medicine,* 46(9), 1193–1203. https://doi.org/10.1016/S0277-9536(97)10048-X

Williams A. 2010. Spiritual therapeutic landscapes and healing: A case study of St. Anne de Beaupre, Quebec, Canada. *Social Science & Medicine*, 70(10), 1633–1640. https://doi.org/10.1016/j.socscimed.2010.01.012

Wilson Edward O. 1984. *Biophilia*, Harvard University Press, Cambridge, MA.

Wilson E. 2008. The nature of human nature. Kellert S., Heerwagen J., Mador M. (Eds.), *Biophilic Design: The Theory, Science, and Practice of Bringing Buildings to Life*. John Wiley & Sons, Inc, Hoboken, NJ pp. 16–35. ISBN 978-0-470-16334-4

Wilson J., et al. 2016. Urban park soundscapes: Association of noise and danger with perceived restoration. *Journal of Park and Recreation Administration*, 34(3), pp. 16–35. https://doi.org/10.18666/JPRA-2016-V34-I3-6927

winnipeg. ca. 2024. 7R Basics, available online: https://guides.wpl.winnipeg.ca/greenchoices/basics, retrieved on 22.08.2024.

Winterbottom D, Wagenfeld A. 2015. *Therapeutic Gardens. Design for Healing Spaces*. Timber Press, Portland.

World's Air Pollution: Real-time Air Quality Index. 2024. Available online: https://waqi.info/, retrieved on 03.09.2024.

World Bank. 2024. What a waste 2.0 A Global Snapshot of Solid Waste Management to 2050, available online: https://datatopics.worldbank.org/what-a-waste/trends_in_solid_waste_management.html, retrieved on 22.08.2024.

World Economic Forum. 2024. Paris is planning to become a '15-minute city', video available online: https://www.weforum.org/videos/paris-is-planning-to-become-a-15-minute-city-897c12513b/, retrieved on 11.04.2024.

World Health Organization, Regional Office for Europe. ⁽1989⁾. *European Charter on Environment and Health*. https://iris.who.int/handle/10665/347390

World Health Organization. 2021, June 11. Cardiovascular Diseases (CVDs), available online: https://www.who.int/news-room/fact-sheets/detail/cardiovascular-diseases-(cvds), retrieved on 23.07.2024.

Wrana J. 2011. Tożsamość miejsca kryterium w projektowaniu architektonicznym. Politechnika Lubelska, Lublin, available online: https://bc.pollub.pl/Content/646/PDF/tozsamosc.pdf, retrieved on 28.08.2024.

Your Guide to Masdar City Central Park. 2024. Available online: https://www.propertyfinder.ae/blog/masdar-city-central-park/, retrieved on 7.03.2024.

Zachariasz A. 2006. *Zieleń jako współczesny czynnik miastotwórczy ze szczególnym uwzględnieniem parków publicznych*, Politechnika Krakowska, Kraków.

Zachariasz A. 2010. *Lyrical beauty, mistery and peace– historic and modern japanese gardens*, Czasopismo Techniczne, 4-A, z. 12, r. 107, Wydawnictwo Politechniki Krakowskiej, Kraków.

Zhang L., Tan P., Diehl J. 2017. A conceptual framework for studying urban green spaces effects on health. *Journal of Urban Ecology*, 3(1), 1–13. https://doi.org/10.1093/jue/jux015

Zhang X., et al. 2020. Indoor air design parameters of air conditioners for mold-prevention and antibacterial in island residential buildings. *International Journal of Environmental Research and Public Health,* 17(19), 7316. https://doi.org/10.3390/ijerph17197316

Zielonko -Jung K., Marchwiński J. 2012. Łączenie *zaawansowanych I tradycyjnych technologii w architekturze proekologicznej*, Oficyna Wydawnicza Politechniki Warszawskiej, Warszawa.

Zimny H. 2005. *Ekologia miasta,* Agencja Reklamowo-Wydawnicza Arkadiusz Grzegorczyk, Warszawa, ISBN 83-89961-30-X

Zube E. 1987. Perceived land use patterns and landscape values. *Landscape Ecology*, 1(1), 37–45. https://doi.org/10.1007/BF02275264

Index

For Product Safety Concerns and Information please contact our EU representative GPSR@taylorandfrancis.com Taylor & Francis Verlag GmbH, Kaufingerstraße 24, 80331 München, Germany

Batch number: 10397794

Printed by Printforce, the Netherlands